Brain Arousal and Information Theory

Brain Arousal and Information Theory

Neural and Genetic Mechanisms

Donald Pfaff

Harvard University Press

Cambridge, Massachusetts and London, England

2006

Library of Congress Cataloging-in-Publication Data

Pfaff, Donald W., 1939–
    Brain arousal and information theory : neural and genetic mechanisms /
    Donald Pfaff.
        p. ; cm.
    Includes bibliographical references and index.
    ISBN 0-674-01920-2 (cloth : alk. paper)
    1. Arousal (Physiology)   2. Information theory.   3. Neurogenetics.
    4. Neurophysiology.   I. Title.
    [DNLM: 1. Arousal—physiology.   2. Nerve Net—physiology.   3. Neurobiology.
    WL 103 P523b 2006]
    QP405.P43 2006
    153—dc22   2005046338

Contents

Preface and Dedication

This book is dedicated to Fred Plum, M.D., University Professor Emeritus of Neurology and former Chair of Neurology at Cornell University School of Medicine. Heroically he devoted his efforts to the understanding and amelioration of disorders of human CNS arousal. He was generous to me in arranging Wednesday lunches with brilliant younger members of his department, Nicholas Schiff, M.D. (Director, Laboratory of Cognitive Neuromodulation and Assistant Professor of Neurology and Neuroscience), Keith Purpura, Ph.D. (Associate Professor of Neuroscience), Jonathan Victor, M.D., Ph.D. (Professor of Neurology), and with his colleague then at Cornell, Michael Posner, Ph.D. Talking with them, I realized that I could expand my own lab's work from explaining the mechanisms of simple instinctive behaviors and specific forms of motivation to more general arousal states. But I had not read about the latter since cross-registering from M.I.T. to Harvard Medical School more than 30 years before. The "pilgrimages" from my Rockefeller lab across the street to Fred Plum's conference room supplied much of the background and courage needed to attempt this book.

I also wish to thank Par Parekh, of The Rockefeller University, for his expert management of the illustrations and tables and extensive textual criticisms.

The book is intended to propose new ideas for my colleagues in this field, but not to substitute for an *Annual Reviews* article in its scientific detail. Therefore, it is heavily referenced to the original literature and to pertinent review papers for proper scholarly support. I have tried to write it in an open and clear fashion, so that readers with college educations in science could understand the main ideas, and appreciate my excitement as a neurobiologist working in this field at this time.

Brain Arousal and Information Theory

1 Toward a Universal Theory of Brain Arousal

For addressing the most fundamental force in the nervous system, we need a precise operational definition. Surprisingly, thinking about arousal's quantitation invokes Shannon's information theory. In humans, loss of arousal is devastating.

Why does an animal or a human being do something under one environmental condition and not another? Why does an animal or human being do anything at all?

Explaining this is going to be difficult. These questions have long been acknowledged as central to neuroscience. But confusion abounds about the very deepest, underlying problem: the elementary arousal of the central nervous system (CNS). I propose, from studying the literature and our own results, that we must explain a function I call "generalized arousal." In the CNS, beneath all of our specific mental functions and particular emotional dispositions, a primitive neuronal system throbs in the brainstem, activating our brains and behaviors. Because this system is universal among all vertebrate animals, the neuroanatomic, neurophysiologic, and neurochemical components of this system are shared among mammals, including humans. The arousal system thus emerging in the human brain drives all of our behavioral responses to stimuli in a manner best understood through the mathematics of information theory. Finally, I will apply our new thinking about this system to our current knowledge of brain mechanisms for sexual arousal and fear.

Arousal mechanisms are exciting and important to understand because at the deepest level they impact all human behaviors. Having evolved over millions of years, their performance determines our earliest responses to environmental stimuli, our expressions of emotion, and our mental health. While Chapters 2–6 on neuroanatomy, neurophysiology, and genetics derive from animal research and Chapter 7 explores engineering concepts, Chapter 8 presents the bottom line. Disrupting arousal mechanisms in humans by accidents, toxins, and so on causes problems ranging from mild loss of vigilance, attention, or sleep to the devastation of the vegetative condition.

◼ What Is to Be Explained? Ethology and the Mechanisms of Arousal

Because arousal is fundamental to all cognition and temperament, its explanation is a holy grail of neuroscience. Consider cognition: You can be aroused without being alert, but not vice versa. In turn, you can be alert without paying attention, but not vice versa. This hierarchical relation persists all the way up to the intellectual function of decision making in the face of uncertainty, the essence of cognitive neuroscience (Fig. 1.1). Likewise, arousal is at the base of all emotional life. You can be aroused without being motivated, but not vice versa. You can be aroused without expressing a strong temperamental feature[1-4], but not vice versa. And so forth, all the way up to subtle variations in moods and feelings that make up our emotional lives. Here is an analogy: For the physics student, emotional behavior can be viewed as a vector. Arousal level determines the amplitude (length of the vector), while the exact feeling and object determine the angle of the vector. Here and below, I contend that melding approaches from the physical sciences with those from the biologic sciences will lead to quantitative behavioral sciences.

Ethological Approaches

When I was a graduate student, the founder and chair of the Department of Brain and Cognitive Sciences at M.I.T., Professor Hans-Lukas Teuber, often told us that experimental analysis of animal behavior is for those who like physics. Ethology, or the study of animal behavior, is for those who like animals, he claimed.

Arousal plays a crucial role in the conceptual foundation of ethology. Satisfying the need for an "energy source" for behavior,[5,6] arousal explains the initiation and persistence of motivated behaviors in a wide variety of species, not just mammals. For example, von Holst[7] used intracranial stimulation to alter drive strength in chickens, while Hinde[8,9] charted the complexities of hormone effects on drive (sex arousal) in canaries. Arousal, fueling drive mechanisms, potentiates behavior, while specific motives and incentives explain why an animal does one thing rather than another.[10] Nor are these findings limited to lower animals. The science of human ethology thrives: Its descriptive, evolutionary thinking will soon be replaced by quantitative, mechanistic explanations.

The *Dictionary of Ethology*[11] not only emphasizes arousal in the context of

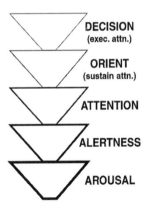
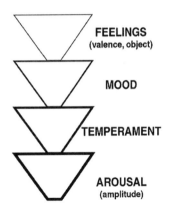

COGNITIVE FUNCTION

DECISION
(exec. attn.)

ORIENT
(sustain attn.)

ATTENTION

ALERTNESS

AROUSAL

EMOTIONAL FUNCTION

FEELINGS
(valence, object)

MOOD

TEMPERAMENT

AROUSAL
(amplitude)

Figure 1.1. Central nervous system arousal is fundamental to all cognitive and emotional functions. Regarding feelings, here is a physical analogy: If emotional behaviors were vectors, arousal would account for their amplitudes, whereas the nature and object of the feeling would account for the angles of the vectors.

the sleep-wake cycle but also refers to the overall state of responsiveness of the animal, as indicated by the intensity of stimulation necessary to trigger a behavioral reaction. Arousal "moves the animal toward readiness for action from a state of inactivity." In the case of a directed action, a founder of ethology, Nikko Tinbergen, would say arousal provides the motoric energy for a "fixed action pattern" in response to a "sign stimulus." The dictionary does not eschew neurophysiology, as it also covers arousal levels indicated by the cortical electroencephalogram (EEG; see Chap. 3).

In this book I take one step beyond classical ethology, which primarily focuses on the careful description of animal behavior. Here I lay bare many of the *mechanisms* for CNS arousal and add a quantitative, *mathematical* approach.

Experimental Analyses of Behavior

Generations of behavioral scientists have both theorized and experimentally confirmed that a concept like arousal is necessary to explain the initiation, strength, and persistence of behavioral responses.[12] Arousal provides the fundamental force that makes animals and humans active and responsive so they will perform instinctive behaviors[6] or learned behaviors directed toward goal objects.[13] The strength of a learned response depends on arousal and drive.[14,15] Hebb[13] saw a state of generalized activation as fundamental to optimal cognitive performance. Duffy[16] goes even further by invoking the concept of "activation" to account for a significant part of an animal's behavior. She anticipated that quantitative physiologic or physical measures would allow a mathematical approach to this aspect of behavioral science (see the following section "A Quantitative Approach to Physical Measurement of Generalized Arousal" and Chaps. 3 and 6). Cannon[12] brought in the autonomic nervous system as a necessary mechanism by which arousal prepares the animal or human for muscular action. Entire theories of emotion were based on the activation of behavior.[12]

Thinking ahead to Chapter 6, "Heightened States of Arousal," note that hormones are effective in raising arousal. Malmo[17] brought all of this material together by citing EEG evidence and physiologic data, which go along with behavioral results in establishing activation and arousal as primary components driving all behavioral mechanisms (pp. 282–301).

This is the classic arousal problem. How do internal and external influences wake up brain and behavior, whether in humans or in other animals, whether in the laboratory or in natural, ethological settings? It is important to reformulate and solve this problem because we are dealing with responsivity to the environment, one of the elementary requirements for animal life. It is also timely to reformulate and solve this problem now because new neurobiologic, genetic, and computational tools have opened up approaches to "behavioral states" that were never possible before. I theorize that explaining arousal will permit us to understand the states of behavior that lie beneath large numbers of specific response mechanisms. Not only is it strategic to accomplish the analysis of many behaviors all at once[18] but also elucidating mechanisms of behavioral states leads to an understanding of mood and temperament. To put it another way, much of twentieth-century neuroscience was directed at explaining the particularity of specific stimulus/response connections. Now we are in a position to reveal mechanisms of entire classes of responses under the name of "state control." Most important are the mechanisms determining the level of arousal as it influences the state of the CNS.

Until now, two conflicts have stood in the way of solving the arousal problem. The first concerns its definition, which I deal with in the following section. The second requires me to explain and then resolve a false dichotomy (see the section "A Quantitative Approach to Physical Measurement of Generalized Arousal").

■ Operational Definition of Arousal

Conflict: historically, the terms "arousal" and "information" suffered similar fates. Both had a vague and slippery character. Everyone knew that arousal exists, for example. It is intuitively obvious, and it is absolutely necessary to explain neurobiologic data. But what exactly *is* arousal? This book dissipates that vagueness, with a concrete operational definition and a quantitative approach to measurement. Claude Shannon's paper on information theory in 1948 performed a similar service for the idea of "information."

Any truly universal definition of arousal must be elementary and fundamental, primitive and undifferentiated, and not derived from higher CNS functions. It cannot be limited by particular, temporary conditions or measures. For example, it cannot be confined to explaining responses to only one stimulus modality. Nor should it be limited to reflex responses to environmental stimuli. Voluntary motor activity and emotional responses also should be included.

Therefore I propose the following as an operational definition that is intuitively satisfying and that will lead to precise quantitative measurements:

"Generalized arousal" is higher in an animal or human being who is: (S) more alert to sensory stimuli of all sorts, and (M) more motorically active, and (E) more reactive emotionally.

This is a concrete definition of the most fundamental force in the nervous system. To understand what we are talking about here, consider two analogies. If we were talking about the geophysics of the planet earth instead of the arousal of the CNS and behavior, this book would be dealing with magma—the hot central core of the earth whose physical distribution controls the magnetic field of the planet. My second analogy is based on primacy in time: If we were talking about the astrophysics of the universe, my book would be dealing with the Big Bang. The primitive arousal responses I discuss comprise the very first, most elementary responses to any sensory stimulus, preparatory for every behavioral response that follows.

All three components—sensory alertness (S), motor activity (M), and emotional reactivity (E)—can be measured with precision. This approach treats behavior as a physical variable to be explained with contemporary neural and genetic techniques. It leads naturally to electrophysiologic recordings of arousal-related responses (Chap. 3).

We arrive at the very same theoretical proposal—that there is a generalized arousal function in the CNS and that it can be defined precisely—by a "reverse approach." That is, we already have our hands on some of the mechanisms underlying arousal, and they lead to the same theoretical idea. Clearly there is a neuroanatomy of generalized arousal (Chap. 2), there are neurons whose firing patterns lead to it (Chap. 3), and genes whose loss disrupts it (Chap. 5). Therefore, the definition of generalized arousal with the reverse approach is as follows: Generalized arousal is the behavioral state produced by arousal pathways, their electrophysiologic mechanisms, and genetic influences. The fact that these mechanisms produce the same sensory alertness (S), motor activity (M), and emotional reactivity (E) as our definition affirms the existence of a generalized arousal function and the accuracy of its operational definition.

While I treat behavior as the most important physical manifestation of arousal, two other types of measures are useful but also potentially confusing. Activation of electrical activity across the cerebral cortex, as measured by the electroencephalogram (EEG) in animals and humans, usually correlates with behavioral activation (Chap. 3), but it is not identical to that activation. Likewise, stimulation of the sympathetic pathways in the autonomic nervous system (ANS) sometimes is correlated with behavioral arousal or, indeed, EEG changes (Chap. 4). We know a lot about EEG and ANS changes and they support our behavioral definition, but they are not precisely identical to arousal as defined. Behavior is the "bottom line."

It may seem mind-boggling that such an arcane and important function as arousal can have such a straightforward working definition. Do not be surprised. Complex behaviors need not have complex explanations. We now understand detailed neuronal, hormonal, and genetic mechanisms for certain mammalian social behaviors,[19,20] for example, even though we do not know, exactly, how locomotion works.

In conclusion, we have a clear operational definition of arousal that relies on physical measures (see the following section) and that is affirmed by the brain mechanisms that lead to arousal (Chaps. 2–4). In the section "The Neurobiology of Arousal Constitutes an Interesting Application of Information Theory" and throughout the book, I make the case that information theory sheds light on how the arousal systems of the brain work.

■ A Quantitative Approach to Physical Measurement of Generalized Arousal

Like many scientific concepts that are widely acknowledged but incompletely understood, arousal neurobiology has generated radically polarized opinions among those working in the field. Early neurophysiologists[21–24] suspected that primitive brainstem systems exerted powerful, monolithic controls over activation of the forebrain. But then the torrents of data being published about neuroanatomic subdivisions of the brainstem, neurophysiologic specificities in reflex controls, and neurochemistries of particular pathways confused some researchers. Several behaviorists—for example, Robbins and Everitt[25]—en-

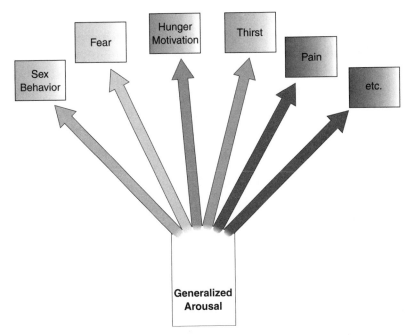

Figure 1.2. This figure complements the equation from Garey et al. (*Proc Natl Acad Sci USA*, 2003; 100:11019). Any given behavior that has a biological motivation depends both on generalized arousal and on a more specific kind of arousal (for example, sexual arousal for sex). Note that the figure allows for interactions among specific arousal states—for example, between sex and pain.

tranced with complex behavioral paradigms, "threw out the baby with the bathwater." Because of the several manifestations of arousal, they claimed, there could not be any elementary, general arousal function. There could not be a Ur-Arousal.

Thus, we have a false dichotomy: either an undifferentiated, singular arousal mechanism *or* no concept at all. I solve this problem by taking a quantitative approach both conceptually and methodologically. The activation of brain and behavior depends on a compound function of a primitive brainstem system common to many states *combined with* neural and hormonal forces that arise from specific biologic needs (Fig. 1.2). The mathematics of arousal is open to investigation.

To move beyond this false dichotomy, I propose a theoretical resolution in an equation that combines both generalized (Ag) and various specific forms of arousal ($As_{1 \text{ to } n}$) such as sex, hunger, and fear.

Eq. 1: $A = F(KgAg + Ks_1As_1 + Ks_2As_2 + Ks_3As_3 \ldots\ldots\ldots\ldots + Ks_nAs_n)$.

where A = arousal, as a function (F) of generalized arousal (Ag) and specific forms of arousal (As). The plus sign is not meant to imply simple linearity, but rather to indicate that A is an increasing function of the variables Ag and $As_{(1\ to\ n)}$, sometimes additive, sometimes multiplicative, and therefore potentially complex. The constants Kg and Ks_{1-n} reflect traits of the individual ("temperament"). The main arousal components (Ag, As) are determined by the situation, the individual's immediate environment[26] (see also Fig. 1.2).

Recent behavioral results show that a generalized force of arousal is operating. We completed several experiments with mice that tapped arousal components S (sensory alertness), M (motor activity), and E (emotional reactivity).[27–30] Then we analyzed the data using a mathematical/statistical tool called principal components analysis (PCA). Generalized arousal accounted for a significant amount of the data—about one-third. PCA uses the mathematical structure of a large dataset—the statistical relations among various response measures—to calculate the contributions of underlying causes. It "lets the subject (in our case, mice) tell us" the structure of its arousal functions. Application of PCA to our five sets of experimental data showed the power of generalized arousal, with the lowest contribution at 29.7% and the highest at 45% of arousal-related data due to a generalized arousal influence.[26] Surprisingly, the overall conclusion that generalized arousal accounted for about one-third of our data held true despite different populations of mice, different investigators, different experimental manipulations and details of response measures, and different configurations of particular factor analysis solutions involving four to six factors for each experiment. Further, the calculation was robust, shown in three ways: (1) The generalized arousal factor was never identical to the first factor of any particular multifactor analysis; and (2) it accounted for significantly more data than in any random-number control or (3) in a stringent control in which marginal averages were held constant but the individual data entries were scrambled randomly. All of these quantitative arguments and the massive amount of neurobiologic data reviewed in subsequent chapters lead to the conclusion that the mathematical structure of arousal functions in the CNS includes a primitive, undifferentiated form I call "generalized arousal."

For influencing generalized arousal, hormones are likely to be useful because their molecular/genetic mechanisms of action are especially well understood. Estrogens, sex behavior, and sexual arousal have already been useful as bridges to the discovery of fundamental arousal mechanisms (Chap. 6). Because estrogens strongly drive female sex behaviors, could the genes coding for their nuclear receptors be involved in a more generalized brain function? To begin exploring genetic influences,[26] we used gene knockouts for the estrogen receptors ER-α and ER-β, two very similar transcription factors and probably gene duplication products. Could functional inactivation (knockout) of the ER-α or ER-β genes influence arousal responses? We studied female mice, individually housed, sleeping in their home cages as they do during the light phase of the daily light cycle. For all sensory modalities tested, α-ERKO female mice were less responsive to sensory stimuli than their wildtype female littermate controls (Fig. 1.3). Surprisingly, disruption of the gene for the closely related ER-β, a likely gene duplication product, did not have the same effect. Likewise, in terms of locomotor activity in running wheels, highest during the dark phase of the daily light cycle, α-ERKO females were less active. Interestingly, this phenotype was dependent on age; older α-ERKO females were subject to the genetic effect, whereas the younger α-ERKO females were not (Fig. 1.4). Again, the differences between β-ERKO females and their wildtype littermate controls were not significant. The graphs for running wheel activity of β-ERKO mice and their controls were virtually identical. In sum, α-ERKO females were less responsive to external stimuli and less motorically active. Quantitative physical measures reveal genetic contributions to arousal.

These mathematical/statistical analyses of behavioral data will resonate with neuroanatomic delineations of massive systems ascending in the mammalian brainstem and affecting crucial basal forebrain neuronal groups (Chap. 2), electrophysiologic demonstrations of multimodally responsive neurons that could serve to alert the animal (or person) to virtually any incoming stimulus (Chap. 3), and ongoing discoveries of genes contributing to arousal-related functions (Chap. 5).

Determining the structure of arousing influences in logical and quantitative terms by using the tools of modern mathematical, neural, genetic, and behavioral technologies will provide a fascinating challenge for the next few years. In fact, a combination of mathematical and experimental approaches to arousal in a genetically tractable mammal such as the mouse will open up the analysis of a CNS function of the most fundamental theoretical and practical significance.

Interactions

Figure 1.2 and Equation 1 both point to the theoretical possibility of quantitative interactions among different types of arousal states. Can heightening one form of arousal affect another? Yes, and relations between hunger and emotional reactivity were among the first to be demonstrated. Campbell and Sheffield[31] reported that food deprivation—increasing arousal due to hunger—caused greater responses to several kinds of stimuli. Relations between pain and sex provide another good example. Both Barfield and Sachs[32] and Antelman[33] showed that pain can heighten sexual arousal. Brown's work[34]

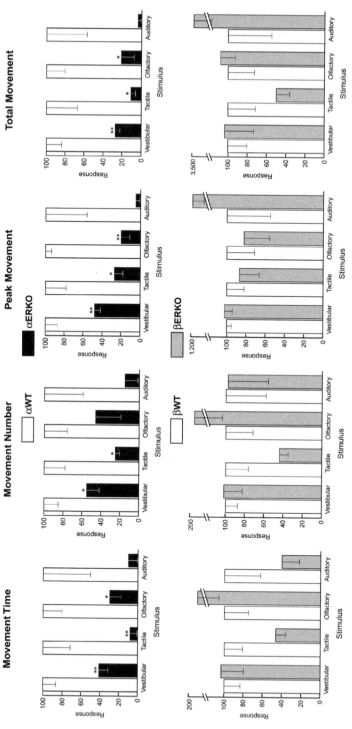

Figure 1.3. Disrupting a gene can alter generalized arousal in female mice. Sensory alertness: Four types of responses by the mice were measured following stimuli in four different sensory modalities. The assay's precision was maximized by testing the mice in their home cages during the part of the daily light cycle when they were quiet. All responses to all sensory modalities were reduced in females whose estrogen receptor-α gene had been disrupted by a genetic technique called homologous recombination (black bars, top panel). They were compared to their wildtype littermate controls (white bars). The perfect genetic control was to disrupt a likely gene duplication product, estrogen receptor-β (grey bars, bottom panel). The bottom panel shows no statistically significant results. (From Garey et al, *Proc Natl Acad Sci USA*, 2003; 100:11019.) Another component of the operational definition of generalized arousal, voluntary motor activity, is shown in Figure 1.4.

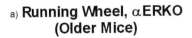

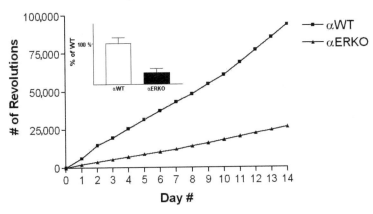

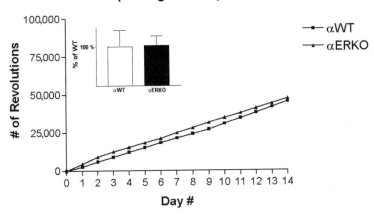

Figure 1.4. Disrupting a gene can alter voluntary motor activity, as a component of generalized arousal, in female mice. These are the same mice as reported in Figure 1.3. In older animals (top panel) but not in younger animals (bottom panel), disrupting the gene coding for estrogen receptor-α significantly reduced the amount of running wheel activity per day. Gene knockout mice (ERKO) were compared to their wildtype littermate controls (WT). We do not know the reason for the effect of age upon the phenotype. As a genetic control, estrogen receptor-β was knocked out in other mice (data not illustrated here) and there was absolutely no effect. (From Garey et al., *Proc Natl Acad Sci USA*, 2003; 100:11019.)

proved that arousing stimuli such as those related to sex can affect fear. Richter[35] deprived female rats of food and saw that their activity levels in response to naturally fluctuating sex hormones were increased. In an ethological setting, Tinbergen[6] showed interactions between sex behavior and aggression, as did Robert Hinde.[9]

Thus, some of the structural dynamics of arousal in animals have begun to be revealed by experiments in which manipulation of one drive state increases another. Evidence for drive interactions such as those shown in Figure 1.2 and Equation 1 has been around for some time. Such interactions justify new thinking about several questions: Are the cellular mechanisms underlying Equation 1 resident within molecular changes (Chap. 4) in the ascending arousal pathways envisioned in Chapter 2? Do the particular types of interactions just mentioned necessarily hold for other drive combinations or other experimental circumstances? Additionally, do conclusions from studies with animals apply to humans? A new field of work will have to address all of these issues.

Math/Stats of General Factors

In retrospect, by directly measuring arousal-related responses with physical methodologies, we have walked past some of the difficulties faced by the psychologists who argue for a generalized intelligence component. A general factor called "g" emerged from a variety of intelligence tests and accounts for 30–37% of the data in the correlation matrices. In fact, in one set of calculations, psychometricians were able to demonstrate that a general factor accounted for about 31% of the intelligence measurement data, even in those tests administered by theorists enamored of selected, specific sets of abilities (see Deary,[36] p. 10). One of the shortcomings suffered by those who have argued against "g" was the tendency to resort to oversimplified either/or formulations. For example, human mental abilities have been portrayed as having at least seven components.[37] However, these separate "frames of mind" have never been proven to be independent, mathematically, of general, core intelligence. It is crucial to understand that the existence and power of a generalized intellectual potential are not vitiated by the simultaneous contributions of specific mental capacities. Moreover, much of the research in that area has been conducted using college undergraduates as subjects. They represent an extremely limited segment among the wide ranges of abilities of human populations. Finally, some IQ theorists have faced political difficulties as they have attempted to extrapolate their thinking to embrace social questions. Generalized arousal theory has no political implications and, instead, is pure science. Medically, the potential therapeutic applications of the brain arousal theories and methodologies impress me with their breadth and are trumpeted in Chapter 8.

In conclusion, we are now in a position to chart the structure of arousal, its dimensions and its mechanisms. Because CNS arousal depends on surprise and unpredictability, its appropriate quantification (surprisingly!) depends on the mathematics of information.

■ The Neurobiology of Arousal Constitutes an Interesting Application of Information Theory

Claude Shannon, born in Petoskey, Michigan, in 1916, was a professor at the Massachusetts Institute of Technology. By the mid-1940s he already had made a reputation working on electrical switching circuits and the early calculating machines. Then he wrote a classic paper[38] that altered the world's approach to all forms of communication systems. What did he do, and why is it important for us?

The concept of information, just like arousal, was intuitively obvious in the common language, but scientifically it seemed vague and slippery. Claude Shannon gave it a precise and useful mathematical definition. By rendering it measurable, he gave scientists a quantitative tool for the description and analysis of every communication device, including the brain. Single-handedly, he created information theory.

Before quoting Shannon's most important equation, I give a commonsense illustration in words: If any event is perfectly regular, say the ticking of a metronome, the next event (the next tick) does not tell us anything new. It has an extremely high probability (p) of occurrence in exactly that time bin. Likewise, in the time bins between ticks, silence has an extremely high probability of occurring. We have no *un*certainty about whether, in any given time bin, the tick will occur. According to Shannon's equation, the information in an event is in inverse proportion to its probability. Put another way, the more uncertain we are about the occurrence of that event, the more information is transmitted, inherently, when it does happen.

What about an array of events that can happen in a given span of time? For simplicity's sake, let's say there are two events, the waving of a red or of a blue flag. If the red flag is waved all the time and the blue flag not at all, no information is transmitted by either flag. And vice versa. However, if the flag selection is random, with a 50:50 chance of either flag being selected, as much information is being transmitted by those flags as can be. When all events in an array of events are equally probable, information is at its top value (see the following example for taste nerves). Disorder maximizes information flow. At the opposite extreme, perfect order minimizes information flow. Coming from thermodynamics, the technical term for disorder in Shannon's equation is entropy. His symbol for entropy is *H*.

Shannon came up with and justified a clear mathematical definition of the content of information in a series of events 1 through *n*.

Eq. 2:
$$H = -K \sum_{i=1}^{n} p_i(\log p_i)$$

The negative sign before the constant *K* is because the log of a fractional probability (*p*) is negative.

Perhaps more attractive is the equation stating that the information content inherent in the event *x* is:

Eq. 3:
$$H(x) = \sum p(x) \log_2 \frac{1}{p(x)}$$

where *p(x)* is the probability of event *x*.

For our purposes, this is the main statement of information theory. Assume we have two possible events with probabilities *p* and (1−*p*). Figure 1.5 shows the curve of the amount of information conveyed. It peaks when the two probabilities are equal—that is, when the observer is least able to predict what will happen, the uncertainty, the surprise, and the *information* content are at their highest.

In the years following the introduction of information theory, biologists rushed to apply the concepts to their problems.[39,40] Of course, it was natural to attempt applications to the CNS and behavior because these involve signaling systems that are quantitative and complex. First, consider the trains of action potentials, or "spikes," exhibited by nerve cells. If time is split up into bins and spikes are counted as either occurring or not within each bin, the amount of information in the spike train is calculated by a formula deriving from Shannon's equation.[41] Similarly, the amount of information inherent in the response to a novel stimulus can be compared to the (smaller) entropy calculated from responses to repeated stimuli.[42]

With respect to neural coding, informatic calculations can be used as statistical tools, measurements of the signaling capacity of sensory systems, and as predictions of cognitive overloads.[43] An entirely new biophysical approach to analyzing neuronal activity to demonstrate a given neuron's maximally informative dimensions depends on Shannon-like calculations.[44] A more specific example is that these calculations can be used to understand the meaning of the firing of dopamine neurons, which obviously respond best to stimuli that represent salient environmental change (look ahead to Fig. 3.5). In this exam-

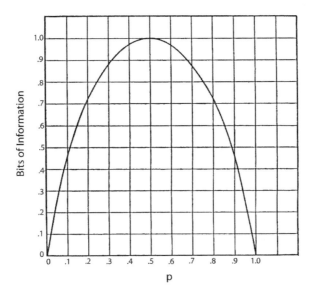

Figure 1.5. Information, by Shannon's calculation, is maximized in situations of greatest uncertainty. Consider a choice between two events, one of which has the probability p and the other $(1.0-p)$. If either one happens all the time (in this figure, the x axis $= 1.0$ or $= 0$), then its informational value is minimized. There are no surprises. On the other hand, if the situation is absolutely unpredictable ($p = .5$), then information is maximized.

ple, the word "salient" means that the stimuli have high inherent information content.

We can also use information theory concepts to see how efficiently a sensory system works. For instance, these concepts tell us how to optimize the efficiency of neural coding. In some cases, such coding would bring information back to the brain's motor control systems for the cybernetic (feedback) control of behavior. In other cases, we are simply talking about sensory systems for their own sake. The distributions of individual taste sensitivities across nerve fibers coming from the tongue into the brainstem of the hamster are statistically independent of each other (from the data of Marion Frank, see Table 1.1) The chance that a given nerve fiber would respond to any given taste submodality (sour, sweet, sour, bitter) was independent of the chance that it would respond to another taste submodality. Such distributions were calculated[45] to maximize information transfer through that taste-coding system. The real was the ideal.

Calculations of uncertainty and unpredictability, which encode the information content of the situation, tell us something about mechanisms of re-

Table 1.1. Informational analysis of responses by single taste nerve fibers to the four basic taste stimuli (salt, hydrochloric acid, sucrose, quinine)

				Information content of response patterns						
Response patterns				Mutually independent responses $R_{NaCl}0\cdot6$, $R_{HCl}0\cdot6$ $R_{Suc}0\cdot4$, $R_{Quin}0\cdot4$		Frank & Pfaffmann data $R_{NaCl}0\cdot6$, $R_{HCl}0\cdot6$ $R_{Suc}0\cdot4$, $R_{Quin}0\cdot4$		Absolutely specific responses $R_{NaCl}0\cdot3$, $R_{HCl}0\cdot3$ $R_{Suc}0\cdot4$, $R_{Quin}0\cdot4$	
NaCl	HCl	Suc.	Quin.	Response pattern probability (p)	Information content $(p\log_2 1/p)$	Response pattern probability (p)	Information content $(p\log_2 1/p)$	Response pattern probability (p)	Information content $(p\log_2 1/p)$
Y	N	N	N	0·092	0·31	0·148	0·41	0·300	0·52
N	Y	N	N	0·092	0·31	0·037	0·18	0·300	0·52
N	N	Y	N	0·040	0·18	0·074	0·28	0·200	0·46
N	N	N	Y	0·040	0·18	0·037	0·18	0·200	0·46
Y	Y	N	N	0·139	0·39	0·111	0·35	0·000	0
Y	N	Y	N	0·061	0·24	0·111	0·35	0·000	0
Y	N	N	Y	0·061	0·24	0·000	0·00	0·000	0
N	Y	Y	N	0·061	0·24	0·111	0·35	0·000	0
N	Y	N	Y	0·061	0·24	0·111	0·35	0·000	0
N	N	Y	Y	0·027	0·13	0·000	0·00	0·000	0
N	Y	Y	Y	0·040	0·18	0·000	0·00	0·000	0

				NaCl	HCl	suc.	quin.				
Y	N	Y	Y	0·040	0·18	0·037	0·18	0·000	0
Y	Y	N	Y	0·092	0·31	0·111	0·35	0·000	0
Y	Y	Y	N	0·092	0·31	0·037	0·18	0·000	0
Y	Y	Y	Y	0·061	0·24	0·074	0·28	0·000	0
			SUM	1·00	3.68	1·00	3.44	1·00	1.96
					bits		bits		bits

NaCl, salt; HCl, hydrochloric acid; suc., sucrose; quin., quinine.

Informational calculations from Pfaff (1975, *Psychoneuroendocrinology*, 1:79–93)

Based on Frank & Pfaffmann data (1969, *Science*, 164:1183–1185) from the glossopharyngeal nerve of the rat and on predictions from mutual independence of responses (optimal cross-fiber pattern code) and absolute dependence of responses (absolutely specific labeled-line code). For each code, probabilities of response to the four individual stimuli are given at the top. For specific response model these were chosen to be in the same ratios as the independent response model, with sum of probabilities = 1·00.

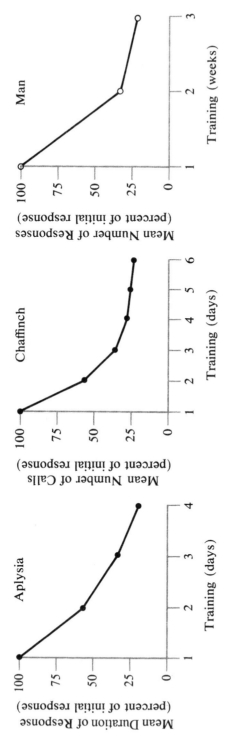

Figure 1.6. Reducing the informational content of a stimulus by simple repetition correspondingly reduces the magnitude of response. Illustrated here by the work and thinking of Eric Kandel (Cellular Basis of Behavior, Freeman, 1976). This phenomenon, a form of neural plasticity called habituation, occurs in a wide variety of animals, including *(left)* a withdrawal response in the mollusc *Aplysia* (data from the Kandel lab); *(middle)* the numbers of calling responses by chaffinches while exposed to an owl day after day (data from Robert Hinde, *Proc Royal Society (B)*, 1954; 142:331); and *(right)* the galvanic skin responses of humans upon repeated presentations of a tone (from Kimmel and Goldstein, *J Exp Psychol*, 1967; 73:401).

ward. The human brain treats rewards differently according to whether they were expected. Activation of a cell group exquisitely involved in reward, the nucleus accumbens, was greater if reward was delayed unpredictably.[46] This difference was also reflected in a functional magnetic resonance imaging (fMRI) study of the striatum, a subcortical motor control system.[47] Interestingly, the left and the right sides of the human prefrontal cortex had different correlations with reward predictability—increasing predictability of reward was positively correlated to activity in the right prefrontal cortex.[48,49] Individual prefrontal cortex neurons recorded in monkeys could be seen to reflect the reward contingency rather than any particular stimulus or response identity—the probability of action/reward combination was the key feature for these cells.[50] Some of these results may be due to subcortical dopamine neurons, long connected with reward mechanisms, because the phasic activation of these neurons in an experiment with two monkeys varied according to probability of reward. The higher the probability, the less activation was seen.[51,52] Apparently, this type of dopamine (DA) neuron will fire when either the presence or the absence of a reward could not have been predicted. Informational calculations will provide a mathematical approach to understanding this type of neuron. In the sensory thalamus, late responses (as opposed to early "phasic" responses) of nonprimary thalamic neurons to stimuli in all modalities peaked just before the delivery of delayed rewards.[53] In all of these experiments, information content determined the brain's response to reward.

I further argue that information theory plays directly into the understanding of one of the most important types of behavioral change, a prominent form of "behavioral plasticity." Let's turn directly to the role of information theory in brain arousal. It is brutally clear that in a monotonous environment, lacking change, a human or an animal will lose arousal and will become less alert. While change, uncertainty, and unpredictability—high-information environments—promote arousal, the lack of these environmental features decreases it. A soldier or sniper becomes less accurate after minutes of waiting for the target, no matter how sophisticated his weapon. In the lab, observations of reduced responses to repetitive stimuli have become formalized and quantified. The phenomenon, called "habituation," occurs in every animal I know of, including a lowly sea snail called Aplysia (Fig. 1.6A), a variety of laboratory animals such as mice and rats (Fig. 1.6B), and humans (Fig. 1.6C). I would reformulate the "habituation problem" from being simply the neurophysiology of a type of learning to being an example of the general neurobiology of informational change.

Reflecting on Figure 1.6 and Equations 1.2 and 1.3, I base my theoretical approach on the idea that, for a lower animal or human to be aroused, there

Table 1.2. Information content determines responses to a wide variety of stimuli. These responses require arousal.

Responses to:		
Novelty	The discontinuous	Important
The unexpected	Uncertainty	Unfamiliar
Surprise	Untraditional	Dissimilar
The unknown*	Changing	Improbable
The new	Inconsistent	Unhabitual
The unpredictable	Unusual	Unprepared
Dissonance	Disordered	Unstable
The infrequent	Irregular	Nonrecurring
Atypical	State change	Nonconforming
Not uniform	Scarce	Unimitative
Singular	Salient	
Incongruity**	Specific	

* As could trigger fear
**As could trigger laughter

must be some change in the environment. If there is change, there must be some uncertainty about the state of the environment. Quantitatively, to the degree that there is uncertainty, predictability is decreased. Given these considerations, we can use Equations 1.2 and 1.3 to state that the less predictable the environment and the greater the entropy, the more information is available. Arousal of brain and behavior, and information calculations, are inseparably united.

Information theory provides a universal, quantitative approach for understanding neuronal and behavioral responses to a wide variety of human situations (Table 1.2). High-information stimuli produce responses that range from stark fear to belly-shaking laughter. Information theory can reveal similarities and differences in behavioral data for which conventional mathematical statistics come up empty.[54] We will be able to predict whether situations provide too much of an informational challenge for individuals. That is, we will be able to optimize work demands for adults and learning situations for children. In this sense, we will be formalizing and explaining the concept of a person's "channel capacity." This term denotes the quantity of information that any transmission system can pass in a given amount of time. It surfaced in George Miller's colorful claim that people could remember only "7 ± 2" items.[55]

Table 1.2 reminds us that unknown, unexpected, disordered, and unusual (high-information) stimuli can produce fear, while the tension, surprise, and incongruity of other sets of high-information stimuli can explain humor. All of the stimulus characteristics listed in Table 1.2 share an opposition to monotonous, regular, predictable situations that kill arousal. All produce and sustain aroused responses. I propose that reconceiving this large family of responses—in terms of their inherent informational content—will enhance the analysis of their mechanisms, be they electrophysiologic (Chap. 3), mathematical (Chap. 7), or genetic (Chap. 5).

What about playful behavior? Jaak Panksepp, at Bowling Green University, is expert at studying brain mechanisms of emotional behaviors in animals and humans.[56] In his opinion, a certain degree of uncertainty, unpredictability, and lack of control—a significant informational content in the game—is an essential element of play. He also emphasizes that there are serious constraints on play. It will occur only in a very friendly context. Any stress or negative motivational state will shut down our playfulness. Likewise, with respect to the degree of uncertainty, there are limits. The range of outcomes must be benign and cannot include disastrous consequences. With these constraints recognized, unpredictability and lack of total control are essential components of a playful situation. If the outcomes are totally known, where is the fun?

Our theoretical approach stating that informational content and novelty dictate arousal responses can also be applied to social situations (Fig. 1.7). As the novelty—the social information content—goes down, the test animal explores a familiar partner less and less. When a new partner is inserted, investigation responses go up. A genetic micronet underlying this behavior has been proposed (Chap. 6).

In addition to being useful in calculating the essential contents of neural coding and behavioral experiments, information theory sheds light on styles of personality and the development of temperament. Jerome Kagan, a professor of psychology at Harvard, has spent most of a lifetime studying child development with an emphasis on the development of different types of personalities and temperaments. He distinguishes "high reactive" infants who show vigorous motor activity and distress to certain environmental stimuli from "low reactive" infants. Infants with the distressed, "inhibited' temperaments grew into individuals who tend to avoid novelty. In contrast, uninhibited infants grow into individuals who welcome novel people and situations.

Recently he has explicitly drawn the concepts of uncertainty and surprise into his arguments.[1] It is particularly the unfamiliarity, the unpredictability, of the stimulus that causes the distress. Kagan draws not only central nervous system structures such as the amygdala (a forebrain region associated with con-

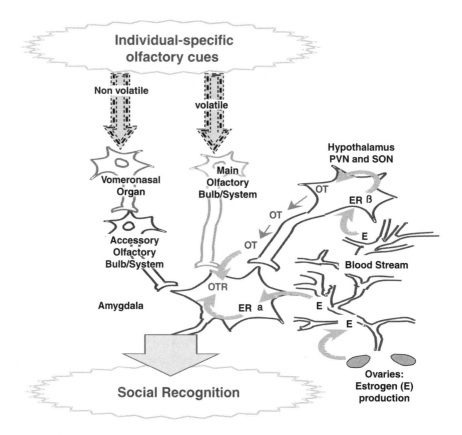

Figure 1.7. A four-gene micronet supporting social recognition in mice. (From Choleris et al., *Proc Natl Acad Sci USA*, 2003; 100(10): 6192–7.) Estrogens circulating in the blood are bound to estrogen receptor (ER)-β in the paraventricular (PVN) and supraoptic nuclei (SON) of the hypothalamus. As a consequence, they increase transcription from the gene coding for oxytocin (OT). Meanwhile, estrogens are bound to estrogen receptor-α in the amygdala. As a consequence, they increase transcription for the gene coding for the oxytocin receptor (OTR). This is important, because from the work of the lab of Thomas Insel, we know that OT in the amygdala fosters social recognition. The neuroanatomy makes sense: mice use pheromones and odors, whose signals bombard the amygdala. The consequences of improved social recognition include increased affiliative behaviors and decreased aggression.

trols over emotion, especially fear) into his thinking but also the autonomic nervous system. A frightened, high-reactive, inhibited kid might be said to have a "reactive" sympathetic nervous system, compared to a calmer, less reactive child. The fact that these data gathered in children can predict temperaments evident in adulthood is perhaps most striking. A scared, inhibited child is more likely to become socially avoidant and worried, compared to a low reactive child who is more likely to become a sociable, calm adult. Kagan and his colleagues have applied their theories of how infant responsiveness to novelty (uncertainty, information) influences temperament to the results of brain scanning.[57] They studied young adults who had been categorized by the authors as either inhibited or uninhibited when they were small children. The former showed greater fMRI responses than the latter in the amygdala.

Throughout all of Kagan's typing of personalities and measurements of autonomic and CNS reactivity, information theory was lurking behind his behavioral measures. High degrees of novelty, uncertainty, and surprise obviously have high information content, which predicts children's responses, be they behavioral, CNS, or autonomic.

Finally, on the cutting edge of medicine, information theory applications can provide sensitive, quantitative, and detailed diagnostic profiles of a variety of fatigue states (e.g., chronic fatigue syndrome, fibromyalgia, Gulf War syndrome) and sleep disorders. For instance, how do we reliably distinguish between a pathologic fatigue and normal sleepiness? The ability of informatic calculations during well-chosen test protocols to reveal unexpected differences or similarities in CNS function and behavior will add dimensions to our diagnoses of fatigue states and sleep disorders and thus guide therapies.

In retrospect, information theory has been lurking behind behavioral investigations and neurophysiologic data all along. First, in clear and simple logic, consider what is required for an animal or human being to rouse himself to action. Second, consider what is required to recognize a familiar stimulus (habituation) and to give special attention to a novel stimulus. Third, from the experimenter's point of view, information theory provides methods for calculating the meaningful content of spike trains and quantifying the cognitive load of certain environmental situations. New questions can be asked: How much distortion of a sensory stimulus field is required for novelty? What kinds of generalization from a specific type of stimulus are allowed for a given type of response? In my field, the information theoretic approach will help us to turn the combination of genetics, neurophysiology, and behavior into a quantitative science. We can use the "mathematics of arousal" to help analyze neurobiologic mechanisms.

In the most general terms, the applications of information theory to neurobiology will help us to turn heaps of genetic and biophysical data into real explanations of how the brain functions. Information theory is an essential component of systems biology. In particular, for this book, it offers us a theoretical entry to the mathematics of arousal.

▪ Claims for This Chapter and Introduction to Chapters Following

So far, we have achieved a rigorous operational definition of generalized arousal and a theoretical equation for predicting arousal level, reported quantitative measures of generalized arousal in a laboratory animal favorable for genetic study, and estimated that generalized arousal accounts for about one-third of the data in experiments with mice. We have also shown that the changes in the stimulus environments of animals and humans, the unpredictabilities and uncertainties that heighten arousal, are measured most precisely by the mathematics of information theory. Therefore, the equations of Claude Shannon and the field he originated do not only summarize neurophysiologic and behavioral results but also suggest new avenues of investigation. Looking forward, both the genetics and the mathematics of the arousal of brain and behavior are open to investigation in coming years.

In sum, the arousal systems of the brain are fundamental to cognition and temperament. Their behavioral manifestations can be measured with the precision of physical variables and their mechanisms investigated with genetic and biophysical techniques.

In this book, the experimental and clinical literature on arousal reviewed in Chapters 2 through 6 lead to the universal theory of arousal presented in Chapter 7. In Chapter 2, I conceptualize a large body of knowledge about the neuroanatomy of arousal-related pathways. As a consequence, I can propose a model in which sparser long-distance connections influence local neural modules. Playing a major role is a crescent of neurons along the ventral and medial borders of the brainstem, which is present in both animals and humans. I assert that the combined activities of these neurons, as they affect cortical arousal and autonomic arousal, constitute the basis of the generalized arousal effects found in the behavioral studies referenced. In other words, the neuroanatomy discussed in Chapter 2 corresponds to the quantitative behavioral results outlined in Chapter 1. Through understanding the electrophysiology (Chap. 3) and some of the genetics (Chap. 5) as well, we have in hand the very mechanisms that themselves argue powerfully for the existence of a generalized arousal function. The biologic substrates prove the concept.

The accumulation of facts and the precision of the hypotheses posed here lend themselves to questions raised frequently by engineers dealing with physical systems. Therefore Chapter 7, in addition to introducing a universal theory of arousal systems, raises some of these questions.

Chapter 8 recounts a few of the applications of this area of scholarship to the understanding of human behavior, normal and abnormal. Because arousal is fundamental to cognition, temperament, and mood, these applications are both crucial and numerous.

I have introduced a system that is universal, natural, and permanent. It underlies the first responses to all stimuli and therefore influences everything that happens thereafter. This system is exciting to study because its phenomena occur fast and they are important for all aspects of human mental and emotional life. It is particularly timely to write on this subject now because of the plethora of new experimental techniques that can be brought to bear on this classic problem. Arousal of brain and behavior—we have defined it and quantified it. Now I will explain it.

2 Anatomy Is Not Destiny, but a Little Neuroanatomy Helps

Many neurochemically distinct pathways with overlapping functions guarantee arousability in the normal individual. Frequently they work by having "long-distance lines" tuning local modules.

The first clues as to how brain arousal systems work can be found in their neuroanatomic structure.

This chapter highlights several theoretically important features. *First*, their multiplicity and redundancy are designed to prevent failure. *Second*, there are primitive old "master cells" among a crescent-shaped region of nerve cells in the lower hindbrain. Their ascending axonal networks alert animal and human brains. These provide the most primitive means of "waking up" the forebrain. Also in the hindbrain, in contrast to the prevailing theory that descending controls over spinal cord are separate from ascending controls over the cerebral cortex, I claim some nerve cell groups in this "arousal crescent" do both. Here, and later in Chapter 7, I face the question, The suprachiasmatic nucleus of the hypothalamus is famous for its biologic clock that manages circadian rhythms, but how does it cooperate with the brainstem's arousal system to time the activation of behavior? *Third*, I propose that limited numbers of long-distance connections work by influencing modules composed of large numbers of local connections.

■ Multiplicity and Redundancy of Ascending Arousal Pathways Prevent Failure

Five major neurochemically distinct systems work together to increase arousal. They use norepinephrine, dopamine, serotonin, acetylcholine, and histamine as transmitters.[58] They all begin in the brainstem and converge in the thalamus (from the Greek *thalamos*, antechamber to the cerebral cortex) *or* in the basal

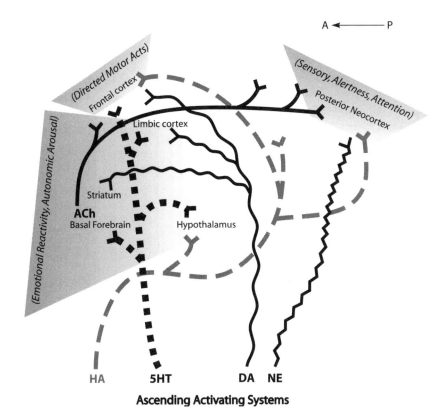

Ascending Activating Systems

Figure 2.1. Simplified schematic representation of some major ascending systems present in animal and human brains that serve to support central nervous system arousal and activate behavior. Four sensory modalities feed these systems in obvious ways: touch (including pain), taste, vestibular, and auditory. Norepinephrine-containing systems (NE, also known as noradrenergic) tend to emphasize projections to the more posterior cerebral cortex (P, except for occipital cortex) and to support sensory alertness. Dopaminergic systems (DA) tend to project more strongly to anterior, frontal cortex (A) and to foster directed motor acts. Serotonergic (5HT) neurons project preferentially to a more ancient form of cortex (limbic cortex) and hypothalamus, and to be involved in emotional behaviors and autonomic controls. Cholinergic neurons (ACh) in the basal forebrain support arousal by their widespread projections across the cerebral cortex. Histamine-producing neurons (HA) likewise have extremely widespread projections that actually originate in the hypothalamus and are strongly associated with increased CNS arousal.

forebrain (Fig. 2.1). They overlap and cooperate. The systems' very multiplicity ensures against failure.

Four sensory systems feed ascending arousal pathways in a straightforward fashion. These clearly show how vestibular, somatosensory, and auditory stimuli as well as taste stimuli on the tongue could arouse an animal or human be-

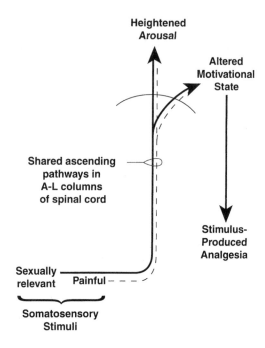

Figure 2.2. Two types of stimuli on the skin, those that are painful and those that are relevant to sexuality, overlap massively as their signals ascend from the spinal cord to the forebrain. Both heighten arousal. Sex motivation is affected, and certain kinds of stimulation cause analgesia in the face of continued painful stimulation. (From Bodnar, Commons, and Pfaff, *Central neural states relating sex and pain*. Baltimore: Johns Hopkins University Press, 2002.)

ing. Pain mechanisms further dramatize how a vastly amplified somatosensory signal from the skin or the viscera could wake up a person quickly. Moreover, pain pathways and sex signals overlap and share the ability to cause states of high arousal (Fig. 2.2). These four sensory systems fit the main proposals of this chapter so nicely that we have illustrated them in Figure 2.1. The two remaining sensory modalities pose special problems.

Smell. Electrical impulses triggered by odor stimuli enter the brain through tracts at the bottom of the basal forebrain, not the brainstem. However, they head directly toward a primary receiving zone that itself is connected with high degrees of arousal, both sex and fear. This is a complex set of nerve cell groups called the amygdala (Chap. 6). In fact, the massive impact of smell on the amygdala itself explains arousal following olfactory inputs.

The amygdala is a primitive forebrain structure and part of the rhinencephalon, also called the "smell brain." Exactly how smell stimuli signaling to the medial and cortical subnuclei of the amygdala excite "output" neurons in the central subnucleus of the amygdala is an interesting question for neurophysiology. The fact that they are open to hormonal influence[59] opens up some possibilities for further thought.

In addition to important influences of smell in the amygdala, some odor-signaling nerve cell fibers make their way to the hypothalamus, triggering electrical impulses there by neurons that send their signals to the ventral midbrain. In doing so, they probably activate cells in the ventral tegmental area (VTA), which has neurons that produce dopamine. In all of these ways smell can activate behavior in an animal or human.

Thus we see that olfactory signals do not have to travel the brainstem arousal pathways to reach forebrain cell groups that alert the animal. They are already there! But even though we can explain how olfactory signals do the job of arousing an animal, an interesting theoretical question remains. Does the different route to arousal taken by smell signals explain their unusual behavioral properties? What about the commanding nature of olfactory signals in lower animals? How about the ability of smell to set off feelings of misery and pain in young adults who suffer multiple chemical sensitivities (MCS) in association with chronic fatigue syndrome? And what do we think about the ability of olfactory auras to predict the onset of schizophrenic episodes?

Visual stimuli. One of the opportunities to explain how visual stimuli get our attention lies in their projection to a midbrain region called the superior colliculi (SC). Visual signals make their way directly to the outer layers of the colliculi (Latin for little hills because these are bumps on the surface of the brainstem at midbrain levels). According to their salience, the signals filter down to deeper levels. In particular, novel visual stimuli impacting the deep layers of the SC trigger orienting responses, including eye movements, toward those stimuli.[60] Some of the SC neurons send neurons up to the thalamus, where they synapse in the pulvinar nucleus and in the reticular and medial cell groups of the thalamus.[61] These in turn alert the cerebral cortex.

The other opportunity to understand the visual/arousal connection originates in the thalamus itself. Part of the thalamus deals specifically with visual signals heading toward the part of the cerebral cortex that does the same. However, some rough features of visual stimulation make their way to the large thalamic region called the pulvinar (which also receives important noradrenergic input), whose axons to the cerebral cortex can command attention by visual receiving zones there.[62] Murray Sherman and his colleagues claim that, in the thalamus, a "bursting" or surge of electrical action potentials permits detection of a new stimulus. This pattern of electrical activity has recently been shown to depend on T-type calcium channels.[63]

A problem for the future is that these connections cannot be the whole story. There is a possibility that, in addition to the bottom-up explanation of visual arousal offered here, there exists a top-down style of control.[62] Visual salience as determined by cortical regions may impose selection criteria on the brainstem regions just mentioned. Such processes happen prior to attentional mechanisms,[64,65] which are not yet understood. Can the convergence of visual signals with arousal pathways explain the primacy of visual life for governing our human behavior?

Throughout the literature on visual neurophysiology, I see the emphasis on novel stimuli, salient stimuli, unexpected stimuli, moving stimuli. All of these qualities speak to the essential informational content of the visual field and thus highlight the importance of an informatic approach (introduced in Chapter 1) to understanding their arousing properties. I predict that firing rates of neurons in these visual systems are perfectly correlated with calculations of the information content of the visual stimuli presented.

Convergence. The important point is that these various sensory signals converge. Whether in the basal forebrain (as the electrophysiologist Barbara Jones and the neuroanatomist Zaborszky would emphasize), or in the nonspecific thalamus (as Mesulam and Steriade would emphasize, from neuroanatomic and electrophysiologic results in the thalamus), a loud, commanding "neuronal shout" must be organized. Then it must be distributed broadly to command the attention of a wide variety of higher-level perceptual processors and motor control cell groups. How the basal forebrain does its job is obvious. Cholinergic and glutaminergic nerve cells there distribute their axons widely across the cerebral cortex.

How the nonspecific thalamus does its job appears more complex. The midline and intralaminar thalamic nuclei receive impressive inputs from the ascending brainstem arousal systems highlighted in this chapter.[66] Even visceral signals are heard.[67] Then the outputs from these nuclei excite the ancient portions of the cerebral cortex as well as the amygdala.[68] Mesulam[69] reviewed succinctly how cell groups in the thalamus that are *not* limited to one sensory modality gather their signals to "shout loudly" around the ancient limbic cortex and other cortical regions, which, similarly, are not limited to one sensory modality. Most exciting is that Mesulam distinguished clearly (p. 48) between thalamic mechanisms limited to processing narrow sensory channels versus those we are discussing, which change the information-processing *state* of the entire cerebral cortex. With precision, Mesulam's neurology matches McCormick's electrophysiology.[70] He and Bal recorded neuronal activity in thalamocortical cells and found two distinct states: rhythmic bursts of activity correlated with sleep, and tonic activity (due to blocking a leak potassium current) during waking states. In this book I am not dealing with point-to-point stimulus-response connections. I am explaining changes in the *states* of the CNS, which in turn govern entire classes of responses.

Neurotransmitter Systems

Having dealt with the special problems of olfactory and visual modalities, let's return to the main theoretical idea, the brainstem's ascending arousal system. Here I summarize the most important properties of the neurochemically distinct brainstem arousal systems in enough detail to match the descriptions of their electrophysiology (Chap. 3) and their genetics (Chap. 5).

I propose that brainstem mechanisms that control arousal of brain and behavior, using the neurotransmitter systems we will deal with now, ascend toward the forebrain following either a "low road" or a "high road." The low road includes ventral pathways, evolutionarily more ancient, impacting (at least) the cholinergic neurons of the basal forebrain. These in turn flood the cerebral cortex with acetylcholine. The high road includes dorsal pathways, more recently evolved among vertebrates and especially well developed among primates, which impact the nonspecific thalamic nuclei—in other words, those nuclei not dedicated to a single stimulus modality.

Norepinephrine (noradrenaline). Noradrenergic axons leave the brainstem for the forebrain traveling a high road (the dorsal noradrenergic bundle) or a low road (the ventral noradrenergic bundle). An extensive body of knowledge has been reviewed in several excellent papers.[58,71–73] Historically, the field was broken open when the histochemical technique of Falck and Hillarp was applied with brilliance by a generation of young Swedish neuroanatomists led by Tomas Hokfelt, Kjell Fuxe, and Annica Dahlstrom. Quickly, they discovered the cells of origin of ascending noradrenergic systems and the axons heading either on the high road toward the thalamus and cerebral cortex or the low road toward the hypothalamus and basal forebrain. In both cases, fields of termination are neuroanatomically diffuse, rather than highly ordered and point to point. Projections to the neocortex are widely interpreted as supporting sensory alertness, while those to the forebrain, including the amygdala,[74] help to control cardiovascular, visceral, and neuroendocrine functions.

Early extensions of the histochemical work into ultrastructural investigations with the electron microscope further revealed the granular vesicles of noradrenergic neurons and laid the morphologic basis for physiologic and chemical studies of this transmitter's release.[75–77] Regarding the dorsal bundle, or the high road, an important source of cell bodies is a small well-contained group of neurons pigmented cerulean blue and called the locus coeruleus (LC). It comprises the sole source of noradrenaline for the newly evolved neocortex. Later in this chapter I consider the broad functional implications of these projections. Briefly, their greater density in the posterior regions of the cortex has been implicated in heightening sensory alertness.[78] Within the LC, some of the neurons contribute axons to virtually all of the projection regions of the CNS, while others specialize by projecting only to certain terminal fields.[79] Regarding the ventral bundle, or the low road, some of the cell bodies lie in cell groups numbered A1 and A2, in the ventrolateral medulla, and in the nucleus of the tractus solitarius. The functions of these projections to the hypothalamus, the basal forebrain, and the amygdala are treated later. While projections to the frontal cortex have been used to explain some of the straightforward, strongly arousing effects of amphetamines,[80] noradrenergic synapses in nucleus accumbens can be drawn into the phenomena of reward.[81]

Special attention is due the LC because of its incredibly wide distribution of noradrenergic axons (reviewed in Foote, Bloom, and Aston-Jones[72] and Rajkowski et al.[82]). The molecular apparatus associated with noradrenergic (NA) transmission is present virtually throughout the brain.[83] NA fibers from LC innervate a large percentage of the primate thalamus. I puzzle, however, why fewer are found in the midline nuclei of the thalamus—a quandary because these may be implicated in the arousing effects of visual stimuli. NA pathways to the cerebral cortex spread out to follow several trajectories. The result is a dense, orderly laminar innervation of lateral neocortex consistent with a strong effect on sensory processing there.

The low road of fibers from LC accounts for the NA innervation of the amygdala, the septum, and the hippocampus, all regions of the evolutionarily ancient limbic system of the forebrain that are of great importance for the control of emotions. LC inputs to the hypothalamus and basal forebrain[84] can easily be tied to NA influences over neuroendocrine phenomena and biologically regulated behaviors. Finally, it has been thoroughly documented that LC fibers reach the spinal cord. This fact feeds the idea that hindbrain cell groups supporting arousal are not simply divided into those that "go up" or "go down." Some do both (Chaps. 5 and 6).

The importance of LC helps us move "beyond specificity." Much of twentieth-century neurobiology, as I noted earlier, was devoted to the analysis of how one particular stimulus but not another could evoke one particular response but not another. LC gives us the neuroanatomic scope, and now we have the genetic and behavioral tools to explain systems such as LC, which govern entire classes of responses through changes in CNS state.

How does LC acquire this power over the state of alertness of an individual? By virtue of its inputs, as well as the outputs I just reviewed. By and large, sensory afferents do not get to LC directly. The very strong inputs that must account for LC's integrative powers are those from the medullary and pontine reticular formation in the hindbrain.[85,86] These collect from a variety of sensory modalities in a generalized fashion and feed the LC as a nodal point in ascending arousal circuitry. Among them, stress-related and autonomic-related inputs must be of special importance.[87,88] Transmitters involved include noradrenaline itself, serotonin, and excitatory amino acids such as glutamate.[89] Importantly, not all the inputs have come from below, ascending toward the forebrain. Some are descending. LC receives axonal projections from a small preoptic area cell group devoted to sleep.[85] Also, indeed, LC receives information (via the dorsomedial hypothalamus) from the "biologic clock" in the hypothalamus, the suprachiasmatic nucleus (SCN).[90] Chapter 7 addresses the question as to whether arousal mechanisms are bipolar—in other words, bidirectional.

Regarding direct studies of arousing functions of LC, I can easily address the low road for NA pathways, those projecting to the hypothalamus and basal forebrain. In the ventromedial hypothalamus, NA plays a crucial role in the initiation of sexual arousal and female sexual behavior. NA inputs to cells in this nucleus excite their electrical activity. Such cells are at the top of the neural circuit for estrogen-dependent female sex behavior[20,91,92] and are found among estrogen-receiving neurons. Now, recording from molecularly identified cells in the ventromedial hypothalamus allows us to study exactly how NA inputs turn on this sex behavior circuit in the female (Chap. 6). In the male, as well, NA projections to the hypothalamus are important. Destroying them affects hypothalamic neuropeptide expression and reduces penile erections.[93] NA cells in the brainstem groups A1 and A2 also project directly to the amygdala,[94] where they can influence mechanisms for the control of fear. NA projections to gonadotropin-releasing hormone (GnRH) neurons have obvious functional importance because the entire reproductive system in both females and males depends on these few hundred neurons in the preoptic area (POA) of the basal forebrain. It is important to appreciate the NA input from the A1 cell group to these neurons[95] because NA is released in the POA when these neurons are activated and, indeed, fosters the GnRH surge that begins the process of ovulation. Likewise, ventral, low-road NA projections from the A2 group to oxytocin neurons in the hypothalamus affect their responses to fear[96] and the responses of hypothalamic corticotropin-releasing hormone (CRH) neurons to immunologic stress.[97]

In sum, the functional consequences of NA actions in the CNS are enormous, both through NA ascending and descending projections (reviewed by several researchers[98–100]). Adrenalin (epinephrine), as well, is centrally involved in stimulating motor activity.[101] Much attention has been given to cognitive effects mediated by NA terminations in the frontal cortex (e.g., Bouret, Kublik, and Sara[102]) and to NA effects on electrical activity elsewhere in the cortex (e.g., Pineda, Foote, and Neville[103]). Equally important are the implications for the management of stress.[104–106] In my lab, we pay particular attention to their neuroanatomic connections to the motoneurons for penile erection and thus male sexual arousal.[107] Finally, descending NA projections help to regulate sympathetic autonomic functions.

Throughout the literature on noradrenaline, researchers emphasize its effects on increasing attention to "salient" sensory stimuli—which are stimuli that are not expected, not predictable, uncertain. Thinking back to our discussion of information theory in Chapter 1, we see the potential use of information theoretic calculations to predict, engineer, and understand how different environments affect arousal. Information theory provides the mathematical content of brain arousal theory.

Dopamine (DA). Dopaminergic axons also course from the brainstem toward the forebrain via two predominant routes. From the substantia nigra, they go forward to innervate a huge noncortical motor control region of the forebrain called the striatum because of its histologic appearance. This forebrain region interests me because current research suggests it may do many things beyond simple motor control. A second DA system arises in a loosely formed cell group in the midbrain called the ventral tegmental area (VTA). These DA neurons are famous for innervating the phylogenetically ancient cortex called the limbic system, which is known to be important for control over motivational states and moods. Notably, some of these DA axons reach the prefrontal cortex,[108] where they synapse on neurons that coordinate the left and right sides of the frontal cortex.[109] This is important in part because the two sides of this region of cortex have opposite effects on arousal and mood: Heightened activity on the right side is associated with lousy feelings in humans, while heightened activity on the left side is associated with good positive feelings. Just as important, some of these prefrontal cortical neurons project back to the VTA.[110] This type of connection emphasizes the bipolar, bidirectional feature of arousal systems highlighted in Chapter 7.

These DA projections can be distinguished functionally from the NA axonal trajectories by DA's tendency to synapse in more anterior regions of the cerebral cortex (Fig. 2.1), those associated with motor activity. This trajectory can be contrasted to more posterior (except for occipital cortex) trajectories of NA axons, associated with sensory processing.

DA clearly contributes to the maintenance of a waking state.[111] When the concentration of DA in the synaptic cleft is elevated by knocking out the gene for the DA transporter protein that would efficiently remove it, the resultant mice have remarkable behavioral hyperactivity in novel environments, making them look like animals on psychostimulants.[112] This phenotype may be due to DA terminations in the shell of a basal forebrain region called nucleus accumbens.[113] Disorders of DA physiology have been claimed to contribute to schizophrenia, and some of the animal pharmacology intended to test this idea has implicated the dopamine D2 receptor.[114]

Among the five genes that code for DA receptors, not all are alike. Most striking are the antagonisms between the biochemical effects of D2 as opposed to D1 receptors. In fact, even the different isoforms of D2 receptor gene products have distinct functions. A long form is mainly postsynaptic and a short form is primarily presynaptic.[115] We are just beginning to understand how these molecular differences serve neurophysiologic functions controlling behavior.

Most exciting for my lab are the findings that DA terminations in the basal forebrain foster sexual arousal. Elaine Hull and her colleagues have made a compelling case that DA release in the preoptic area fuels sexual arousal and pursuit of females by males.[116,117] In females as well, dopaminergic function is important. Cheryl Frye[118,119] has shown that progesterone metabolites acting in VTA can foster female sexual behaviors. Clearly, some of the important VTA actions on female sexual arousal are in the basal forebrain, because Jill Becker reported sex hormone actions on dopamine-sensitive systems there.[120] These experiments showed strong hormonal and DA effects on movement control, including the kinds of locomotion involved in female courtship behaviors. Chapter 6 enlarges on these themes; sexual arousal puts forth a well-understood set of mechanisms illustrating how arousal can work. In the words of Professor Susan Iversen at Oxford,[121–123] the DA systems are "integral to motivational arousal."

I propose a significant revision of DA theory as I consider the projections of VTA neurons to the nucleus accumbens. This nucleus has obvious connections to the phenomenon of reward. As a result, DA terminations there were interpreted merely as coding reward. I think things are actually a bit more complicated and subtle. A strong line of research from Jon Horvitz[124] at Boston College demonstrates that the salience of stimuli *per se* is the critical requirement for activating DA neurons (illustrated in Chap. 3) rather than reward *per se*. Put another way, the reward value of a stimulus is just one way for that stimulus to gain salience. Destroying DA systems markedly slows responses to salient stimuli and leads to the omissions of responses.[125] Continuing on this theme, DA neurons are not necessarily sensitive to reward itself but instead seem to signal anticipations and predictions of future rewarding events[126] (cf. Wilson et al.[127]). Fluctuations of dopamine levels in nucleus accumbens during rewarded acts[128] are consistent with the following new point of view: DA projections to nucleus accumbens signal excitement and arousal, not simply reward. Further, two prominent features of DA neurobiology tie its arousing power to information theory. First, the *salience* of a stimulus depends on unpredictable and uncertain change, which maximizes information transfer (Chap. 1). Second, it is the variation in *predictability* of a reward that engages DA systems. Thus, informatic calculations provide the concepts and the metric fundamental to DA and arousal neurobiology.

Serotonin. The serotonin system, also, was revealed by the Karolinska group using histochemical procedures pioneered by Falck and Hillarp. Serotonin neurons are found in the raphe (derived from Greek, fence) nuclei of the brainstem; the nuclei are so named because of their long, tall, skinny appearances on the midline of the brainstem. The impressive ascending projections of these neurons[129] drench the forebrain in serotonin. Serotonin-transporting axons travel toward the forebrain via a high road (dorsal trajectories) and a low road (ventral trajectories) (as reviewed by Jacobs and Azmitia[130]). Most of these fibers ascend from the dorsal raphe nucleus. A more ventral group includes fibers from both the dorsal and the median raphe nuclei of the midbrain. Dense serotonergic terminals are seen in several regions of the limbic system, including the hippocampus, septum, and amygdala (cartooned in Fig. 2.1), and the hypothalamus. In addition, serotonergic projections to the thalamus are strongest in limbic-associated thalamic nuclei.[130] The hypothalamic projections have been confirmed with a new and precise neuroanatomic method that uses choleratoxin.[131]

Raphe/serotonergic connections to cortex are widespread, to olfactory cortex, to the olfactory bulb itself, and to the more recently evolved neocortex. Gene products coding for serotonergic receptors (Chap. 5) are equally widespread. The diversity of molecular structures and expression of the fourteen genes coding for serotonin receptors tells us that the logic of their cellular functions will be quite complex. For example, while serotonin can depolarize neocortical cells and enhance their excitability consistent with neocortical arousal,[132] other mechanisms also operate. Recording from dissociated prefrontal cortical pyramidal neurons, Carr and Surmeier[133] found that serotonin could inhibit a sodium current in a manner dependent on phospholipase C and protein kinase C, which of course would reduce excitability. How separate receptor subtypes contribute to different electrophysiologic effects remains to be worked out.

How does the serotonin system gain its physiologic power? Part of the answer lies in its inputs. Nerve cells in the raphe nuclei receive significant afferents from adrenergic neurons in the lower brainstem[134] as well as from neurons expressing peptides related to arousal such as orexin.[135,136] Theoretically important are inputs descending from the forebrain, showing the bipolar nature of arousal systems (Chap. 7). These include influences from the limbic system and hypothalamus,[137] and even from the pineal gland, which imposes a daily rhythm on serotonin production.[138]

The functional importance of ascending serotonin-bearing axons is undeniable. Working both directly on the cerebral cortex and indirectly through the basal forebrain cholinergic neurons,[139–141] serotonin affects the balance between waking states and sleeping states. Serotonergic agonists activate the EEG by suppressing low-frequency activity.[142] Antagonists at serotonin receptors block cerebral activation.[143–145] Evidence from the use of gene knockout mice suggests that 5-HT1B receptors might be especially important.[146] Electrical stimulation of serotonin neurons activates low-amplitude high-frequency EEG. Lesions reduce it (reviewed in Dringenberg and Vanderwolf[147]). Some of these functions serving generalized arousal of the forebrain also bear on sexual arousal. A large percentage of dorsal raphe cells express the gene for the sex hormone receptor estrogen receptor-β (ER-β).[148] Through serotonergic receptors in the hypothalamus, these raphe neurons can affect female sex responses such as lordosis behavior.[149,150,151,152] Thus, generalized arousal contributes to a specific form of arousal (Chap. 6).

Acetylcholine (ACh). Nerve cells important for arousal and transmitting through the use of acetylcholine are found in two major regions: the dorsolateral nucleus of the tegmentum (DLT), in the brainstem at the border between the pons and midbrain, and in the basal forebrain. The latter are spread out across the magnocellular preoptic nucleus, the diagonal bands, and the septum (reviewed in Jones[153]). They project widely across the cerebral cortex. They receive their integrative power from the generous inputs of glutamatergic, cholinergic, noradrenergic, dopaminergic, and histaminergic ascending arousal system neurons that inform them (e.g., Jones and Cuello[139] and Hajszan and Zaborszky[154]). Their activation produces an EEG consistent with alertness and increasing waking behaviors.[155] The former group of ACh neurons, in the DLT, project into the thalamus,[156] the hypothalamus, and the basal forebrain. Activation of DLT neurons renders thalamic neurons more sensitive to sensory information.[157] Their stimulation activates the cortical EEG. Their destruction bilaterally in a human being, with neighboring areas damaged following mechanical or vascular accidents, produces a vegetative or comatose state. By the combined actions of these two cholinergic cell groups, ACh affects behaviors associated with arousal through both cortical and limbic activation (reviewed in Jones[158]).

How do these ACh projections work to produce arousal? In the cerebral cortex, ACh synapses work both through receptors classed as nicotinic[159] and those classed as muscarinic.[160] The electrical consequences of these projections are just now beginning to be worked out. For example, due to the occupation of these receptors, cerebral cortical cells can be excited by inward-flowing calcium currents and by the reduction of outward-flowing potassium currents. But these are just the beginning of the story. Still unknown is the meaning of having two widely separated cell groups—the DLT and the basal forebrain—producing ACh, with both crucial for arousal.

Histamine (HA). Histamine clearly influences arousal and the sleep-wake cycle. Consider, for example, that taking antihistamines makes us sleepy. All of the histamine-producing neurons are located in the posterior hypothalamus in small groups collectively called the tuberomammillary nucleus (TMN). These neurons send diffuse, widespread projections to many brain areas, including the neocortex, and are more active in association with behavioral arousal.[161,162] They can also excite cholinergic neurons in the nucleus basalis of the basal forebrain, by acting through H1 and H2 receptors there.[163] In turn, these cholinergic neurons help to wake up the cerebral cortex.

Behaviorally, an H1 receptor blocker can reduce responsivity of female mice to external stimuli[164] (look forward to Fig. 5.1). Males were less affected. Electrophysiologically, as well, the histaminergic system strengthens CNS transmission of afferent sensory signaling.[165] HA neurons receive part of their physiologic powers as a result of their multiple inputs from brainstem arousal systems.[153] Moreover, their reciprocal innervation with the sleep-associated ventrolateral POA should help to enforce rapid and powerful state changes between sleep and wakefulness.[166]

Histamine signaling has many influences on the sleep-wake cycle, and our understanding of the mechanisms involved is incomplete. We do know that HA can depolarize neurons, making them more able to respond to afferent information;[132,167] but we know little about H2 and H3 receptor actions compared to the large amount of data on H1 receptors. Finally, wiping out the HA by itself does not put the animal to sleep.[142,168] This underlines a main point of this section: Redundancy among neurochemical mechanisms serving arousal prevents system failure.

Lessons

I have summarized and simplified sixty years of neuroanatomy to highlight major features of arousal pathways. This summary leads to four outstanding characteristics of arousal system neuroanatomy.

Table 2.1. Ascending arousal systems have both generalized and specific components

General features of arousal systems

1. Anatomical connections are not point-to-point.
2. Arousal effects are not limited to individual stimulus/response combinations. Instead, they are prerequisites for the occurrence of particular stimulus/response combinations. They change CNS state.
3. There is considerable redundancy, neurochemically and functionally, across arousal components.

Particular features of arousal systems

1. Cell bodies of NE, DA, 5-HT, ACh, and His systems are different, so inputs are not identical.
2. Axonal distributions are different, so target regions are not identical.
3. The greatest functional impacts are not identical among NE, DA, 5-HT, ACh, and His.

NE, norepinephrine; DA, dopamine; 5-HT, serotonin; ACh, acetycholine; His, histamine.
Modified from Bodnar et al., 2002.

First, the separate origins and neurochemistries of the different pathways should allow some to survive and function even when others are damaged (see also Chap. 7). The fact that all these systems can respond to many forms of stimuli and distribute their excitation widely fits them well in serving the common goal of brain arousal.[69,70,132,139,153,158,169,170]

Second, even as these pathways have generalized features, they are not identical (Bodnar, Commons, and Pfaff[18]; see Table 2.1). Their favorite regions of termination in the cerebral cortex or basal forebrain and their greatest points of functional impact provide opportunities for separate manipulation. As mentioned, noradrenergic terminals are somewhat denser in the posterior regions of the cerebral cortex (except occipital) than the anterior. Correspondingly—with those cortical regions devoted more to sensory processing than to motor acts—noradrenergic pharmacology has been tied most closely to sustained alertness.[78,125,171–173] In contrast, dopaminergic pathways travel more anteriorly toward the forebrain, terminating in regions convincingly tied to motor control. Correspondingly, their neuropharmacology reveals them as crucial for motor acts directed toward salient stimuli.[124] The emotional valences of different ascending systems may also differ: NE signaling more stressful stimuli, and DA closer to positive rewards. In turn, serotonergic function has been implicated strongly in the control of mood. Witness the popularity of SSRIs—selective serotonin reuptake inhibitors—for treating depression. The challenge for molecular pharmacology is to figure out how the biophysics of multiple receptor subtypes plays into the neurophysiologic and behavioral functions I have just mentioned. There are at least fourteen genes coding for serotonin receptors. What do they all do? Just as striking is the evidence that individual dopaminergic receptor subtypes—D1 versus D2—have opposite biochemical effects within the same neuron. Calculating how the differential contributions of all of these histochemical systems and their separate receptor subtypes provides an exciting challenge necessary for a full understanding of how arousal works (see, e.g., Jones[153,158]).

Third, even though I have emphasized pathways that ascend toward the forebrain, some very important controls over arousal begin in the basal forebrain and descend (Fig. 2.3). An excellent example is the small group of γ-aminobutyric acid (GABA) neurons in the ventrolateral preoptic area (vlPOA). Clifford Saper and his lab at Harvard Medical School[166,174,175] have used a wide variety of experimental approaches to demonstrate that these neurons are important for normal sleep to occur and, further, that they work by innervating some important cell groups in the ascending arousal systems. Within preoptic/hypothalamic circuitry, Saper may have identified a "sleep switch."[176] We are testing this theory by using viral vectors overexpressing the genes for glutamic acid decarboxylase 65 and 67 (GAD 65; GAD 67) microinjected into the vlPOA and assaying generalized arousal, as defined in Chapter 1. Another example is the suprachiasmatic nucleus (SCN) of the hypothalamus, discovered as a biologic clock in mammals by lesion studies.[177,178] Now, the "non-image-forming visual system" emanating from the SCN is understood to influence many biologic rhythms, including daily changes in arousal (see reviews[179–181,182,183]). Looking forward, the genetics of our CNS biologic clocks is introduced in Chapter 5 (compare Young[184] and Young and Kay[185]).

Antonio Damasio talks about the importance of understanding the interplay between these descending and ascending influences in his 1999 book *The Feeling of What Happens*.[186] How is the unity of the body maintained during changes in arousal states when many major neurophysiologic forces are thrusting upwards while other important influences are descending in the brainstem and spinal cord? One interesting group of neurons in the paraventricular nucleus (PVN) of the hypothalamus may help answer this question. These cells have axons descending from the hypothalamus, not only to alter EEG-controlling mechanisms in the brainstem but also, importantly, to control autonomic nervous system function. Some of these neurons deposit CRH—corticotropin-releasing hormone—at the top of the stress hormone hierarchy. Others release arginine vasopressin (AVP), which conserves body fluids in the wounded or thirsty animal, as far away as the lower brainstem or even the spinal cord. In addition to these long descending arcs, there are other ladder-like descents, for

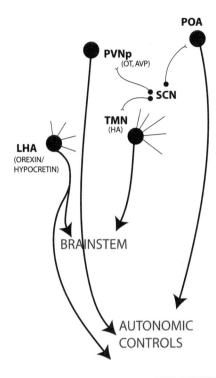

Figure 2.3. Arousal is controlled top-down as well as bottom-up. Lateral hypothalamic area (LHA) orexin neurons project to monoamine-expressing cell groups in the lower brainstem and even to the spinal cord. Neurons that express oxytocin (OT) and arginine vasopressin (AVP) in the parvocellular portion of the paraventricular hypothalamic nucleus (PVNp) control autonomic arousal through the lower brainstem and spinal cord and affect EEG arousal through projections to locus coeruleus. Hypothalamic neurons containing histamine (HA)in the tuberomammillary nucleus (TMN) have widespread projections and receive inputs from a biological clock, the suprachiasmatic nucleus (SCN). Preoptic area (POA) neurons have descending axons that affect sleep and autonomic physiology.

example, from the hypothalamus to the central grey of the midbrain, connecting with midbrain to ventromedial medulla (hindbrain).

All of these mechanisms control autonomic aspects of arousal: heart rate, respiration, peripheral resistance to blood flow, gut physiology, and skin temperature. Arthur Loewy and his group at the Washington University School of Medicine in Saint Louis have used transneuronal labeling by pseudorabies virus[187] as well as other neuroanatomic techniques[188,189] to demonstrate the cell groups giving rise to descending autonomic controls (see also Chaps. 5 and 6). It is exciting that, for the control of skeletal muscle tension associated with sexual arousal[190-194] and locomotion,[195] broad-ranging long-axon neuronal influences have been discovered.

Arousal pathways go both up and down. This fact raises theoretical questions about the bidirectional or bipolar nature of controls over arousal, an important part of my theory introduced in Chapter 7.

Finally, the very absence of correlations among the activities of the various converging systems summarized here leads to two of their most important virtues: (1) If information passage into the forebrain is to be maximized, a *lack* of correlation, an *unpredictability* of peak activity among the ascending systems, according to information theory, is required; and (2) it is very good for the health of arousal systems that the separate ascending systems *not* be correlated. That is precisely what is needed for stable performance. When one subsystem is down, another can do the job. These features are important because, historically, a lack of perfect correlation was used as an argument against the arousal concept. Instead, the lack of correlations among the systems exemplifies precisely what is needed for high-information, biologically adaptive performance.

Having summarized a lot of neuroanatomy and derived at least four lessons, I nevertheless admit that the "heavy lifting" still lies ahead. One of our major tasks in the rest of the book is to examine how the ultrastructural and biophysical details of synaptology in these systems, converging in the thalamus, the cortex, or the basal forebrain, manage the physiologic jobs outlined in Chapter 1. For now, however, let us double back to an important source of arousal. Many of the cell bodies of a primitive, core system lie in a crescent of "master cells" deep in the brainstem.

■ Primitive "Master Cells" in the Brainstem Provide a Neuroanatomic Core that Theoretically Matches the Behavioral Data

If the previous section presents a theorem, this section describes a corollary. This corollary deals with the most primitive origins of arousal and answers the question, "Where does it all begin?" I propose that the oldest, most powerful cells for waking up the brain are those forming a roughly crescent-shaped domain of cells and axons, an arousal crescent as important for the generation of brain arousal as the Fertile Crescent was for human civilization. The arousal crescent, an abstract idea, is physically formed by a broad ridge of cells near the ventral border of the brain, its bottom, curving into a ridge of cells along the midline. Figure 2.4 provides a three-dimensional abstract rendering of this concept. The crescent is formed both on the left and the right sides of the brain equally. It stretches all the way from the back of the hindbrain, where the brain turns into the spinal cord, forward to the flexure of the brainstem, and onward toward the front of the brainstem where the midbrain turns into the thalamus and hypothalamus. In fact, two of my teachers at M.I.T., the neuroanatomist Walle J. H. Nauta and the neurophysiologist Patrick Wall, felt that, all the way from the lower spinal cord to the forebrain, a reticular core of cells could be distinguished from discrete, specific sensory and motor processing modules. The reticular core accounts for the alerting and emotional properties of our reactions to stimuli from the external world.

These are the master cells, the brainstem neurons most primitive and powerful in providing the mechanisms for elementary arousal of brain and be-

Figure 2.4. An extended system of reticular neurons distributed bilaterally roughly as a crescent along the bottom and medial edges of the lower brainstem support CNS arousal. The most posterior picture is at the bottom of the page, and we proceed forward (anterior) through three sections to the topmost picture, which is in the midbrain. Arousal-related neurons (cartooned as large black dots) are also found in locus coeruleus (second from top), the dorsolateral nucleus of the tegmentum, and the raphe nuclei (topmost section). Concept is superimposed on a brain atlas, which for rats is from Paxinos and Watson and for mice is from Franklin and Paxinos.

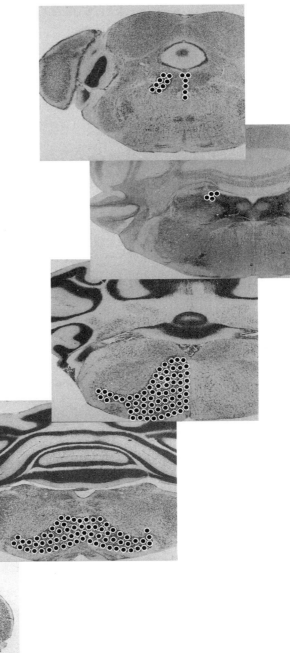

havior. They explain the ability to respond promptly and alertly to stimuli with a significant content of information.

The general, abstract neuroanatomic features of the arousal crescent are conserved from the brains of lower mammals like mice into the human brain. Indeed, it is almost embarrassing for the human CNS that the neuroanatomy of these master cells in the brainstem is virtually identical to that of the cells in the mouse brainstem. I propose that for elementary arousal of brain and behavior, these master cells are the oldest and the most important.

Inputs to these master cells show their tremendous integrative power, proving their qualification for the job of alerting the brain to a wide variety of bodily conditions and environmental situations (see, for example, Van Bockstaele, Pieribone, and Aston-Jones[196]). These include stress,[197] sexual sensations,[19,107,198] vestibular[199] and painful stimuli,[200–202] thermal sensations,[203] signals from the viscera and blood,[135,189, 204–207] and hormonal inputs.[95,208,209] In addition to this tremendous range of emotionally significant inputs, the master cells of the arousal crescent gain integrative power from the interesting geometry of their dendrites. "Dendrite" comes from the word for "tree" in Greek, and a large percentage of a nerve cell's information comes from synapses on its dendrites. Our master cells differ from specific sensory signaling pathways of the CNS in which the first dendritic segment of a nerve cell—as we go out from the cell body—is longer than the second segment, which is longer than the third, and so on. However, in our arousal-related reticular neurons,[210] the first dendritic segment is *shorter* than the second, which is *shorter* than the third, and so on (Fig. 2.5). This geometry vastly increases a neuron's ability to gather significant information from a wide range of inputs.

The outputs of many arousal-driving neurons are, likewise, very impressive. We can see how our master nerve cells there become as important for arousal as they are. Three of their features begin to tell us how they gain their influence.

First, consider the structures of the neurons themselves. These are tremendous cell bodies, capable of metabolically supporting huge axonal trees. The great Spanish neuroanatomist Valverde showed that their axons, after a short lateral traverse from the cell, branch into a huge descending tract and a huge ascending tract. These are neurons that you have to respect for their integrative power up and down the neuraxis. The illustration of these cells in Figure 2.6 is based on the work of Valverde, the Scheibels, and the Russian neuroanatomist Leomintov. Primitive neurons, deep in the brainstem, have the power to influence both cortical arousal and autonomic arousal.

Second, even though some of their projections are long, arousal systems also have the capacity to form an ascending "ladder" of connections, thus pro-

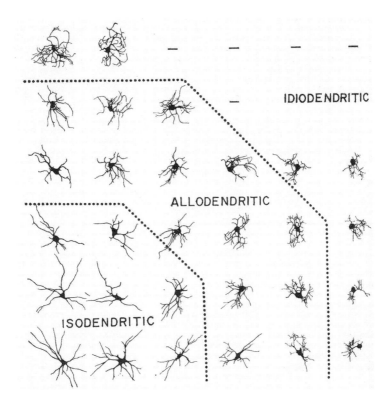

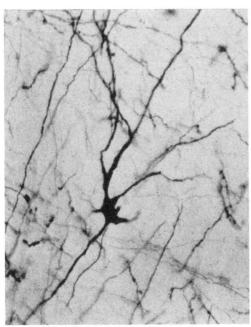

Figure 2.5. *Top:* Isodendritic neurons, conceptualized and illustrated by Ramon-Moliner and Nauta (*J Comp Neurol,* 1966; 126: 311), include properties typical of my hypothesized master cells. The dendritic segments closest to the cell body are *shorter than* the next dendritic segment, which is *shorter than* the next. Isodendritic neurons can be found, for example, in the brainstem reticular formation. In contrast, allodendritic neurons have the opposite arrangement: the dendritic segment closest to the cell body is the longest. They are more typical of specific sensory nuclei. Idiodendritic neurons are not important for this argument. *Bottom:* An isodendritic neuron in the reticular formation, at higher magnification. Isodendritic neurons are well equipped for gathering signals from large numbers of arousing stimuli.

In the top figure: IDIODENDRITIC, ALLODENDRITIC, ISODENDRITIC

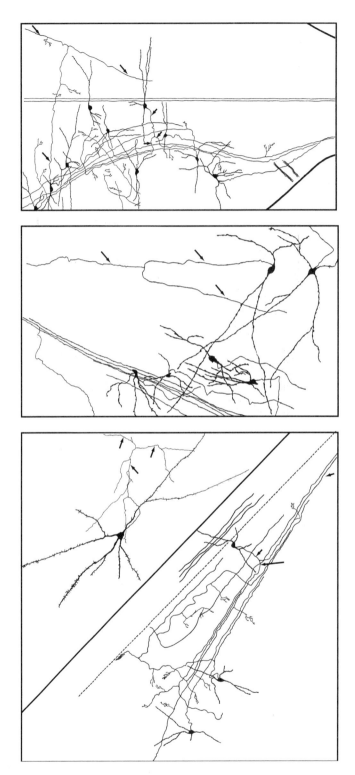

Figure 2.6. Five examples of large reticular formation nerve cells whose trunk line axons bifurcate into ascending and descending limbs. Such neurons would be among my theoretical master cells—primitive neurons influencing arousal going both toward the hypothalamus and cerebral cortex and toward the spinal cord. For each of the five cells illustrated, the bifurcating axonal limbs are denoted by three small arrows. Illustrations of neurons photocopied from Valverde (*J Comp Neurol*, 1961; 116:71), Leontovich and Zhukova (*J Comp Neurol*, 1963; 121:347), and Valverde (*J Comp Neurol*, 1961; 117:189 and 1962; 119:25).

viding the capacity for amplification of an arousing signal as it is transmitted toward the forebrain. For example, ventral medullary neurons in the arousal crescent project up to the locus coeruleus [211] and up to the nucleus of the solitary tract.[212,213] Similarly, cells in the reticular nucleus paragigantocellularis send axons to locus coeruleus, thus to have their own signals amplified by LC noradrenergic influences.[213,214]

Third, the importance of the arousal crescent is not limited to its connections that ascend toward the forebrain. Even as some of the axons ascend toward the thalamus, hypothalamus, and basal forebrain—thus to influence behavioral alerting responses and neuroendocrine phenomena—others descend toward the spinal cord. Here they can have important influences over the autonomic nervous systems, both sympathetic and parasympathetic. Arthur Loewy and his colleagues at the Washington University School of Medicine have used modern neuroanatomic techniques to demonstrate brainstem neurons that project to sympathetic premotor neurons as well as vagal preganglionic neurons.[187,215–219] In fact, some of the very same neurons that project to the spinal cord also impact autonomic-related neurons in the lower brainstem.[220] In sum, these brainstem regions carry out descending as well as ascending controls of arousal (for more on this theme, see Chap. 4).

With these three properties clearly understood and more likely to be discovered, what can be said about the range of physiologic outputs from the arousal crescent? Among the activities of neurons in the rostral ventrolateral medulla are central roles in the reduction of pain (reviewed in Bodnar, Commons, and Pfaff[18] and in Mason[203]) and in thermoregulation.[203] Likewise, these neurons are important for sex behaviors,[18,19] and for the crucial postural and locomotor reflexes (e.g., Loy et al.[195] and Schepens and Drew[221]) that are components of a large number of behaviors. Neurons in the arousal crescent participate in respiratory reflexes[222,223] and, indeed, in many reflexes mediated by the cranial nerves (reviewed by Saper[224]). The functions served by these arousal-driving cells boggle the mind and are certainly underestimated by these few examples.

The large neurons deep in the reticular formation of the brainstem that I describe as master cells of arousal systems are phylogenetically ancient neurons. They have been serving behavioral activation throughout the lives of all vertebrate animals. Correspondingly, the amino acid transmitters—glutamic acid as an excitatory transmitter, GABA as inhibitory—are part of a phylogenetically old neurochemical system. It is fitting, therefore, that a substantial number of these reticular cells use glutamate as their transmitter.[153] They have the appropriate synthetic and re-uptake capacities for glutamate to do the job.[225–227] Further, Rasmussen and his colleagues found that blocking glutamate transmission disrupts these cells' neurochemical effects in the basal forebrain. Putting all these arguments together, I submit that the oldest and potentially the most powerful among the long-axoned arousing brainstem pathways are the large master cells in the brainstem reticular formation, transmitting via the release of glutamate.

Finally, I assert that the influences these primitive, powerful brainstem neurons exert through their widespread outputs correspond to the behavioral measures of generalized arousal (Chap. 1). Remember, about one-third of arousal-related responses are due to a generalized arousal factor. These can be accounted for by the old arousal crescent nerve cells. Brain mechanisms for generalized arousal tendencies depend on these master cells. Primitive, powerful hindbrain and midbrain neurons generate primitive, generalized arousal responses.

■ Long-Distance Lines Tuning Local Modules

I am convinced that the most important arousal networks in the brain work similarly to friendship networks. Each of us has our tight-knit group of friends with whom we interact often. But suppose there is one person in our group whose associations branch over to another, completely different, distant group. We would say that, from that second group, we have "two degrees of separation," and so on, until we reach the famous formulation that each of us can relate to any other human with only six degrees of separation. A few long-distance connections tap into large numbers of closely woven groups of local connections. How can we think about this setup systematically and mathematically?

Albert-Laszlo Barabasi, the Hofman Professor of Physics at Notre Dame University, has proposed a theoretical summary of networks in biology.[228] I believe this theory applies to neuronal pathways governing brain arousal. Here is how Barabasi describes the distribution of communication links. Large numbers of people (or neurons, or websites, or biochemical molecules or whatever in biology) have small numbers of connections to others; but a small and predictable number of people (or neurons, or websites, or biochemical molecules or whatever in biology) have large numbers.

We can visualize how this mathematical distribution looks in a systematic way. Figure 2.7 illustrates the theory that the number of persons (or neurons and so on) with a certain number (n) of connections declines rapidly as a function of (n) according to an equation called a power law. This type of lawful decline apparently applies to the number of links on a webpage,[229] the number of citations to physics papers,[230] metabolic networks,[231] and a variety of other ap-

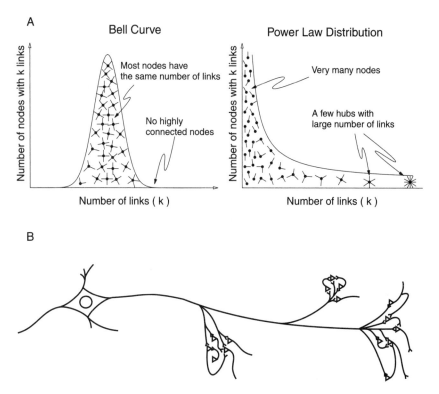

Figure 2.7. (A) Barabasi's contrast of a normal bell curve distribution of connectivities (*left*) versus a situation in which a small number of elements have large numbers of connections (*right*) (from *Linked: the new science of networks*. Cambridge, Mass.: Perseus, 2002, p. 71). If my idea of primitive master cells underlying generalized arousal is correct, they would exemplify these latter types of elements. (B) Following on from Barabasi's ideas, a schematic diagram of a highly connected large arousal-promoting neuron influencing substantial numbers of smaller neurons in modules. Each of those small neurons has smaller numbers of local connections.

plications.[232,233] The effect of a power law type of distribution has been pointed out graphically by Steven Strogatz[234] (p. 255). If the power law equation has an exponent of 2.2, there will be very few highly connected people (or neurons) compared to the less connected ones. People (or neurons) with 10X more links would be 158X less likely. Looking at it the other way around, the addition of even a small number of highly connected people (or neurons, and so on) drastically decreases the average separation between people.[235]

The same idea applies to my arousal system theory. During conversations with Floyd Bloom and Steve Henricksen at the Scripps Institute in California, I realized that an important feature of neurons serving generalized arousal would be to send long axons that influence local modules elsewhere in the CNS (Fig. 2.7, bottom). Certainly, this idea applies to long-distance projections from norepinephrine neurons in the locus coeruleus to the forebrain and from serotonergic neurons in the raphe nuclei to the limbic system. Surely, in their projections to the hypothalamus, such systems find their local modules. Using electron microscopy applied to different hypothalamic cell groups[236,237] with a method pioneered by Laszlo Zaborszky, we found that a high percentage—about two-thirds—of the synapses arose from neurons that were extremely close to the neuron contacted and certainly within the same local cell group. Less expected was that a transmitter famous for being inhibitory in the CNS might participate in this type of system. Steve Henricksen and his colleagues[238] found that GABA neurons in the ventral tegmental area (VTA)—at the very front of the arousal crescent—had fast-firing electrical activity during the onset of movements. Their activity correlated with behavioral arousal. In contrast, the neurons decreased their activity by 53% during sleep. Henricksen proposes that the wide dynamic range of activity of VTA GABA neurons might serve them well to project to the basal forebrain and impact "local-business" short-axoned GABA neurons there, thereby disinhibiting cholinergic neurons,[239] which wake up the cerebral cortex. Once again, a long-distance connection controls a local module.

Henricksen's approach resonates with the theory in this book in at least two other ways as well. First, GABA phylogenetically is a very old transmitter. Given that the arousal crescent is a very primitive system for waking up the brain, it is reasonable to expect that a very primitive transmitter such as GABA should participate. The real surprise here is that certain interesting GABA neurons (in the VTA, according to the Henricksen lab results) are highly connected and project all the way to the basal forebrain and even to the cerebral cortex.[240]

Second, anesthetics acting at the GABA-type A receptor play prominent roles[241] in medicine. Reversing this point, because many anesthetics and hypnotics use chemicals that act on GABA type A receptors, they implicate this transmitter in neurochemical systems mediating arousal (e.g., Tobler et al.[242]). See Chapter 8 for a discussion of the importance of arousal theory for a rational approach to mechanisms of anesthesia.

The arousal network structure hypothesized here protects against many kinds of damage to its neurons. In contrast, loss of key, highly connected neurons in the arousal crescent would markedly decrease the arousability of brain and behavior.

Recounting in detail the powerful neuroanatomic pathways of arousal permits us to think about the "reverse" approach to the definition and physiology of arousal: Arousal is as the arousal crescent does. *Arousal is that set of functions that arousal mechanisms produce!* The neuroanatomy, the biophysics, and the genomics of those neurons tell us how they produce what amounts to the operational definition of arousal (Chap. 1). Experimentally, we can use discoveries about the biology of those neurons to fine-tune our behavioral assays. In turn, those experiments refine our thinking about the neurons in question. These approaches converge. They will lead to a mature science of the arousal of brain and behavior in which cellular mechanisms and behavioral outputs are in perfect accord.

■ Summary and Hypothetical Implications for Human Behavior

I have painted the neuroanatomic picture of arousal pathways with a broad brush in order to illustrate these principles:

- Multiple neurochemical systems provide the redundancy the CNS needs to ensure against disastrous failure. The characteristics of these systems overlap but are not identical.

- A primitive core of master cells in an arousal crescent in the brainstem represents the obvious correlate of generalized arousal measured in behavioral tests. The primitive core is present in the human brain as well as in the brains of lower animals.

- Arousal neuroanatomy shows us that relatively small numbers of long-axoned connections can fine-tune local modules of neurons.

- Different parts of the arousal pathways do not need to be correlated with each other. In fact, the very absence of correlation helps to maximize information flow in the CNS.

Other authors have pictured for us how far this kind of neurobiology can bring us toward the explanations of the higher aspects of human behavior. We can begin to see how the components of arousal systems combine so that new, collective behavioral phenomena will emerge. Clearly, the noradrenergic synapses widespread in the neocortex of primates[243] have behavioral effects similar to those of amphetamine (e.g., Berridge and Stalnaker[80]). These underlie sustained attention, or vigilance.[78,171] Long-lasting personality dispositions with respect to reactivity and arousability go under the names "temperaments"[244-248] and "moods."

Some of the implications for human behavior have more to do with frank medical questions. At the present time, surgical anesthesia is not a medical sci-

Table 2.2A. Major components of human consciousness and their clinical disturbances

Some clinical functions

    Activation of rostral mechanisms of arousal, psychological state functions, and cognitive integration

    Generation and regulation of psychological states of affect, mood, attention, and cathexis

    Generation and integration into consciousness relatively focal higher psychological functions of perception, memory, learned motor acts, and anticipation

Some clinical disturbances

    Disorders of arousal

        Coma

        Stupor

        Sleep disorders: hypersomnia, insomnia, narcolepsy

    Disorders of attention

        Distractibility, inattention, locked-on vigilance, obsessiveness

    Disorders of affect or emotion

        Anxiety or panic

        Agitation, irritability, lack of restraint, logorrhea, aggression

        Apathy, akinesia, mutism

    Disorders of psychic energy

        Indifference

        Fatigue syndrome and its congeners

    Disorders of global cognitive function

        Delirium and fugue states

        Multimodal dementia

        Vegetative state

    Impairments of focal conscious properties

        Agnosia

        Apraxia

        Aphasia

        Loss of anticipation

        Amnesia

Source: Adapted from Plum, F., N. Schiff, U. Ribary, and R. Llinas, Coordinated expression in chronically unconscious persons. *Philos Trans R Soc Lond B Biol Sci,* 1998. 353: pp. 1929–1933.

ence and is not quantitative. It is an art. This is unfortunate because, while bad advice from a general practitioner may allow us to die within months and bad surgery can kill us in hours, incorrect anesthesia can kill us in minutes. Notably, the indices used by the practitioner for depth of anesthesia[249,250] are very close to my operational definition of arousal: awareness of sensory stimuli, motoric activity, analgesia, and autonomic responsiveness. Arousal is the exact

Table 2.2B. Clinical criteria for patients in a vegetative state

Time durations
  One month, if persistent more than one year, almost always permanent
    No cognition: absence of consistent responses to linguistic, symbolic, or
      mimetic instruction
    No semantically meaningful sounds or goal-directed movements
    No sustained head-ocular pursuit activity
Functions usually or often preserved
  Brain stem and autonomically controlled visceral functions
    Homeothermia, osmolar homeostasis, breathing, circulation,
      gastrointestinal functions
    Pupillary and oculovestibular reflexes
    Brief, inconsistent shifting of head or eyes toward new sounds or sights
    Smiles, tears, or rage reactions
    Reflex postural responses to noxious stimuli

Source: Adapted from Plum, F., Coma and related generalized disturbances of the human conscious state, in *Cerebral Cortex*, A. Peters, ed. (New York: Plenum Publishing, 1991), pp. 359–425.

opposite of anesthesia. Therefore, as we understand one, we understand the other.

Other medical implications derive from problems with arousal systems. Mesulam,[69] emphasizing the brain's control over entire classes of responses—changes in *state* of the CNS—talks about "confusional states": distractibility, incoherence, inability to carry out a sequence of goal-directed movements (p. 129). Heartrending are the even more severe disabilities associated with vegetative states, stupor, and coma.[251,252] The comatose patient is not aware of his environment and cannot be aroused by stimulation. The stuporous patient requires continuous stimulation to maintain even minimal levels of arousal. The vegetative state preserves autonomic functions but lacks cognition (Table 2.2A). Bilateral damage—for example, by vascular accident or mechanical insult—to the arousal pathways, especially in the brainstem at the level of the upper pons or the midbrain, has disastrous results in the comatose, stuporous, or vegetative state. How these should be distinguished mechanistically leads to important new questions for arousal science. Not all of the clinical disturbances of human consciousness originate in disorders of arousal pathways (Table 2.2B), but those that do are serious and block meaningful human existence.

More abstract and indirect are the contributions of arousal mechanisms to human functions associated with reward and addiction. Clearly, the neuroanatomy is in place. Dopaminergic projections to the shell of nucleus accumbens,[81] for example, show one kind of ascending arousal-related projection to a basal forebrain cell group proven necessary for behavioral reinforcement. How normal reward mechanisms can get twisted into addictive states represents a complex of questions that are still not understood despite decades of study. Further, the ascending arousal pathways documented earlier in this chapter feed the frontal cortex, a cortical region that is important for the appreciation and maintenance of emotional state.[254] Human values may live here (Edelman and Tononi,[255] p. 87), nurtured by noradrenergic, serotonergic, and other arousal-related inputs.

Antonio Damasio, VanAllen Distinguished Professor of Neurology at the University of Iowa College of Medicine, has been the most articulate at explaining how our brainstem arousal pathways play into our sense of ourselves. In *The Feeling of What Happens*,[186] Damasio claims that damage to the ancient arousal pathways hurts our "proto-selves," that is, our core consciousness (pp. 270–271; see also Parvizi and Damasio[256]). Other approaches to human consciousness emphasize the importance of midbrain and pontine arousal pathways in maintaining cortical and behavioral wakefulness,[224] and possible top-down controls called "re-entrant signaling" by Gerald Edelman.[257] Somewhat more concrete would be findings about arousal systems associated with hypnotizability,[258] autonomic changes in meditative states, and arousal-related contributions to brain regions implicated in the appreciation of pleasurable emotions.[259]

Of course, it is a very long scientific road from our logically tight, physically measurable definition of behavioral arousal in small, genetically tractable mice to these major questions of human health and emotion. Therefore, some of the medical issues and items related to public health will have to be revisited in Chapter 8.

3 Arousal Is Signaled by Electrical Discharges in a System of Nerve Cells

Some nerve cells in the brainstem respond to a remarkably wide range of stimuli. They are primitive neurons, having been with us "from fish to philosopher."

Certain nerve cells in the brainstem of the waking animal respond to a wide variety of stimuli. Many of these neurons fit into my arousal crescent concept, extending from the lower brainstem all the way up to the hypothalamus. Rather than having a narrow response spectrum, these nerve cells are multimodal in their responses. Here I review their physiologic characteristics. We will follow the responses of these multimodal-responding neurons up the brainstem and include dopamine (DA) neurons, serotonin (5HT) neurons, and even hypothalamic neurons responding to histamine (HA). The electrical activities of these neurons always depend on unpredictability and uncertainty—the informational content of the stimulus and its environment.

This chapter traces the consequences of activation of these brainstem neurons, both in the electrical activity of the forebrain—for example, the cortical electroencephalogram (EEG)—and for functions like sex and sleep. In humans, substantial bilateral damage to these nerve cells and their projection regions results in a comatose state or, at the very best, a vegetative condition.

The dynamic responses of these multimodal master cells result in part from their neuroanatomy (Chap. 2) and must involve patterns of gene expression (see Chap. 5). The data in this chapter and throughout this book reverse the twentieth-century emphasis on "specificity." Now we have the experimental techniques and theoretical power to try something much more ambitious. We will explain entire states of the CNS and will be able to predict changes of state. These possibilities are exciting because a generalized state of the CNS helps us to understand entire classes of behavioral responses at one time. Generalized arousal states of the CNS match human behavioral dispositions and thus offer great explanatory power.

■ Traveling Up the Brainstem

In the arousal crescent itself (Chap. 2), large numbers of cells respond to wide varieties of alerting stimuli, not just a narrow, specific type of stimulus. This is especially true during recordings from animals carefully prepared to record without the depressing effects of anesthesia. For example, Leung and Mason,[260,261] recording from cells in the medullary raphe nuclei and magnocellular reticular nuclei, ventral and medial in the medulla, discovered neurons that they ordinarily would have classified as pain-related neurons but that also responded clearly to non-noxious stimulation (from the Mason lab[260,261]; see Fig. 3.1). In a much different experimental paradigm, Bruce Lindsey and his colleagues at the University of South Florida recorded from neurons in the arousal crescent that are closely connected with respiration. The neurons responded not only to painful stimuli but also to manipulations of blood pressure and blood chemistry.[262,263] Likewise, Hubscher and Johnson[206] found medullary reticular cells that could respond to inputs from the skin over the entire body. Other medullary neurons receive inputs from a range of somatic and visceral afferents related to decreases in blood pressure.[264] In sum, these and other master cells in the arousal crescent fit the bill perfectly as cells that feed information from many convergent systems to LC and other arousal-related brainstem nuclei. In her *Annual Reviews* article, Peggy Mason theorizes that this broad range of arousal information is used by neurons in the medullary raphe and reticular formation to adjust both behavioral and autonomic responses appropriately to the behavioral state and environmental context of the animal or human.

Moving anteriorly in the hindbrain reticular formation, electrophysiologists can find multimodal neurons and interpret their activity in terms of how they bring information together for particular integrative jobs. Consider vestibular stimuli, which signal changes in head movement and position. It is easy to see why they compel attention. For example, if a monkey is asleep in a tree and begins to lose his balance, he has to rouse immediately and regain his branch. Barry Peterson at Northwestern University was mainly interested in specific vestibular-initiated reflexes; but some of the reticular cells he recorded could respond to quite a variety of inputs. These were multimodal cells.[265,266] Jim Phillips at the University of Washington shows how brainstem reticular neurons gather information for the purpose of controlling eye movements. Some of his neurons, called OPN neurons, change their rate of firing according to levels of consciousness and, when they are firing at a high rate, can be correlated with arousal and visual attention.[267–269] They respond not only to visual and auditory stimuli but also, under optimal conditions, to vestibular and

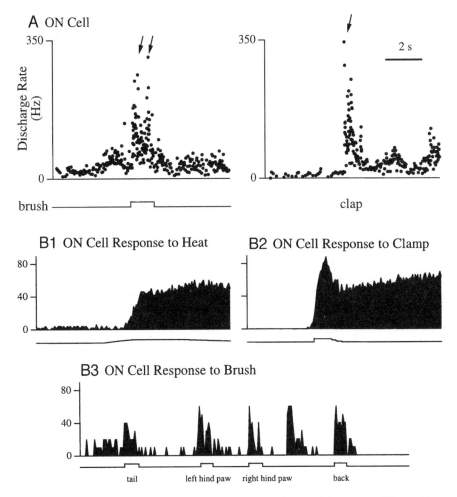

Figure 3.1. Multimodal brainstem neurons, as recorded in the laboratory of Peggy Mason at the University of Chicago. In the terminology of her field, ON cells respond to painful stimuli with increases in discharge rate. They also can respond (A) to a light, innocuous cutaneous stimulus (*left*) and an auditory stimulus (*right*). (B) Another cell that responds to painful heat (B1) or clamp (B2) can also respond to gentle tactile stimuli on many parts of the body (B3). (From the data of Leung and Mason, *J Neurophysiol*, 1998; 80:1630–46 and 1999; 81:584–95.)

somatosensory stimuli. Trevor Drew, in Montreal, wants to know how brainstem reticular neurons control dynamic postural adjustments during cat locomotion. He has found cells whose electrical discharges were stimulated by inputs from all four limbs.[270] Far from having narrowly constricted sensory fields, the neurons Drew studied covered a significant fraction of the cat's body (Fig. 3.2). In all these cases, as we travel up the brainstem reticular formation,

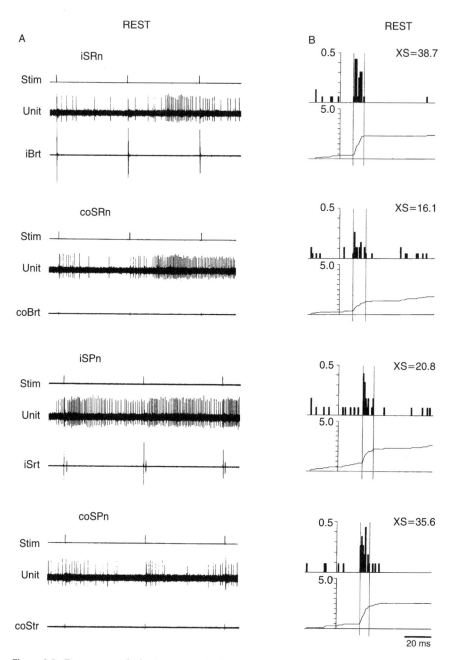

Figure 3.2. Responses of a brainstem reticular neuron to stimuli coming from limb nerves of all four limbs of the body. Receptive fields were not narrowly constrained. Neurons were recorded in Trevor Drew's lab at the University of Montreal, with the animal at rest. Raw data on the left (A). Histograms showing the increased firing rates following stimulation, on the right (B). (From Drew, Cabana and Rossignol, *Exp Brain Res*, 1996; 111:153–68.)

neuroscientists have found cells that serve very well to activate the animal's brain following a wide variety of stimuli.

According to my theory, these multimodal master cells, responding to a wide variety of stimuli and preparing the individual for rapid action, have been with us for a very long time in evolutionary history.[271–274] Deep in the brainstem of the earliest vertebrate animal groups, fish, there are large cells that respond to cutaneous stimuli on one side of the body as well as to auditory stimuli,[275] stimulation of the lateral line organ, and vestibular and visual stimuli (Fig. 3.3). Their axons then cross the midline and descend to the spinal cord, synapsing there on motoneurons that rapidly cause muscle contraction, thus propelling the fish away from the strong stimulus.[276–278] I believe that Nature did not "throw away" these crucial brainstem neurons, but instead has maintained them and their functions throughout vertebrate evolution.

Chapter 2 introduces the far-flung norepinephrine (NE) pathways emanating from locus coeruleus (LC), a cell group in the lower brainstem of all mammalian animals and human beings. But what tells these cells to fire electrical signals at a high rate? After all it is their electrical action potentials that deposit noradrenaline (NA) throughout the forebrain (reviewed, Berridge and Waterhouse[71]). LC cells respond strongly when the informational content of the incoming signal is high: when it is sudden (as opposed to long and steady), salient (unpredicted), and when it is important. To make this last point, Gary Aston-Jones and his colleagues showed that the LC neuronal responses in the monkey brain correspond with the motivational importance of a stimulus and are not purely sensory or motor in their nature.[279,280] That team was able to correlate strong LC responses to meaningful stimuli with good performance of learned behavioral responses. Figure 3.4 provides an example of LC neuronal responsivity. The ability of such a neuron to respond to salient stimuli can be understood by thinking back to the neuroanatomy discussion in Chapter 2. LC neurons receive very strong inputs from master cells in the arousal crescent— for example, the large ("gigantocellular") reticular formation neurons in the ventrolateral rostral medulla.[214]

LC neurons can also integrate longer-lasting states of arousal with the stimulus immediately presented.[281] They fire at higher rates during EEG states of arousal as well as on presentation of exciting stimuli such as a preferred food. How might the arousal state dependency occur? There are at least three chemical mechanisms to think about, based on current data. First, orexin/hypocretin, a gene product associated with behavioral arousal (Chap. 5), can activate LC neurons, probably by decreased potassium conductances.[282,283] Second, LC neurons are exquisitely responsive to corticotropin-releasing hormone (CRH), a hypothalamic neuropeptide essential for responses to stress.[284]

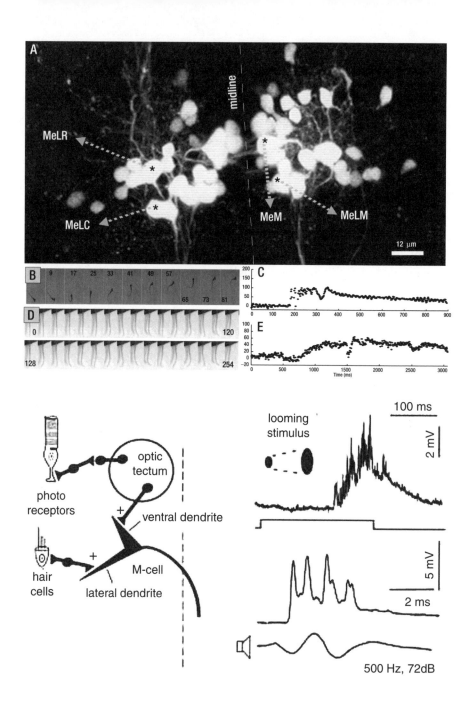

The magnitude of response to CRH depends on the genetic background of the animal.[285] Conversely, neurochemical agents that should decrease arousal in the animal likewise decrease LC electrical activity. Third, the inhibitory transmitter GABA significantly reduces electrical discharges by LC neurons.[286,287] This route of action by GABA probably participates in mechanisms of anesthesia (see Chap. 8). All of these LC response tendencies should, theoretically, be amplified by the expression of connexin proteins, which are known to permit the fast spread of electrical excitation within LC. We are investigating this possibility of rapid amplification of LC signals using patch clamp recording from genetically engineered LC cells. In sum, increased neuronal activity by LC cells results in increased arousal, alertness, and attention.[288]

As we move forward toward the upper brainstem, we encounter the neurons that manufacture and use DA. Recall that dopamine neurons in the ventral tegmental area of the midbrain projecting to nucleus accumbens were once thought to fire exclusively in association with the presentation of a reward.

Figure 3.3. Master cells controlling rapid responses to arousing stimuli were already present in the hindbrain reticular formation of fish.

Above: (A) Naga Sankrithi and Donald O'Malley have labeled these large reticular formation neurons with a fluorescent dye. They are near the midline (in part of what I call an arousal crescent), next to a fiber pathway called the medial longitudinal fasciculus (giving rise to the abbreviation MeL). (From the unpublished work of Donald O'Malley; for example, see Gahtan et al., *J Neurophysiol*, 2002; 87:608, and O'Malley et al., *Methods*, 2003; 30:49.) (B) High-speed recording of the escape response, complete within 81 milliseconds. As documented by Budick and O'Malley (*J Exp Biol*, 2000; 203:2565), a touch to the head triggers a C-shaped bend, a turn and a burst of swimming. (C) Biophysical documentation of increased calcium concentrations in some of the cells shown in panel A during the response to touch illustrated in panel B. (D) Visual stimuli also evoke swimming. Here the tail movements are shown over a period of 254 milliseconds, during a step increase in illumination. (E) O'Malley and his collaborators here document increased calcium concentrations in one of the neurons shown in panel A during visually evoked swimming as shown in panel D.

Below: In the left half of this figure, T. Preuss and D. Faber sketch the pathway by which alarming visual stimuli can trigger electrical activity in a large reticulospinal cell (Mauthner cell, M-cell), which then initiates an escape response. The upper-right panel shows an intracellular recording from such a neuron to a looming visual stimulus, as would be seen in the expanding shadow of a predator approaching.

In the lower half of this figure, they illustrate a direct pathway from the ear to this reticulospinal cell. The lower-right portion of the figure, with a much faster time scale, shows an intracellular recording from this reticulospinal neuron as it responds to an abrupt sound click (representative of a suddenly arousing loud noise). These M-cells have been identified as such in fish and amphibia. (From the work of Donald Faber, Albert Einstein College of Medicine, and his coworkers. See Faber, Korn and Lin, *Brain Behav Evol*, 1991; 37:286; Zottoli and Faber, *The Neuroscientist*, 2000; 6:26; and Canfield, *Brain Behav Evol*, 2002; 61:148.)

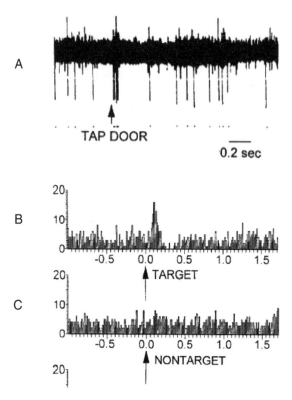

Figure 3.4. From recordings of electrical activity in individual neurons of the monkey's locus coeruleus by Gary Aston-Jones and his coworkers (*J Neurosci*, 1994; 14:4467–80). These neurons respond to novel, unexpected, and salient stimuli. (A) A burst of action potentials stimulated by an unexpected tap on the door of the animal's cage. (B) Increased rate of firing (Y axis) following presentation of the target stimulus during a vigilance task, but not (C) following presentation of a nontarget stimulus.

Now we know that it is not so simple. A newer and more sophisticated theory claims that DA cell firing is not restricted in this way. Instead, DA neurons respond to a much larger category of stimuli that are, for several reasons, salient in the environment.[124] Such stimuli could be salient by virtue of their physical energy, their connection with reward or punishment, or—as suggested by our information theoretic approach—salient by virtue of their low probability and unpredictability. Electrical recordings from multimodal DA neurons support this view[289] (Fig. 3.5). Mechanisms that may serve dopaminergic transmission of arousing stimuli to the forebrain include, for example, DA gating of glutamatergic signals.[290] An alternative possibility has been construed during electrophysiologic experiments with genetically altered mice. According to

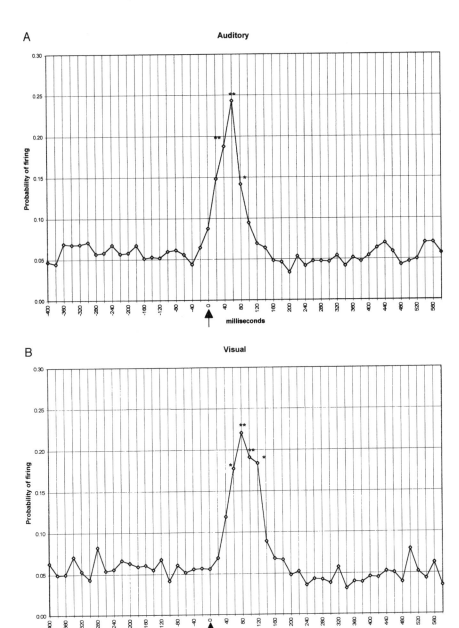

Figure 3.5. Jon Horvitz, from Boston College, has recorded from dopaminergic neurons that will respond to any salient stimulus. His recordings here show increases in firing rate to an auditory stimulus and a visual stimulus. His concept of dopamine neurons' responsiveness encompasses the older theory linking dopamine to expectation of reward. That is, one way of making a stimulus salient is to pair it with reward, but that certainly is not the only way. The dependence of dopamine neuronal excitability upon unpredictability, salience, and surprise fits in perfectly with the application of information theory to arousal systems. Reference and further details in text.

this idea, DA feedback controls the bursts of activity shown by DA neurons.[291] The result is an association between DA firing and the initiation of motor responses directed toward salient stimuli;[292] some of these responses are sometimes termed "appetitive behaviors."[293]

How can this new way of viewing DA neurons be reconciled with older reward-related data? At least two possibilities come to mind. First, it is possible, in keeping with our theme of information theory and arousal, that fluctuations in entropy itself (Chap. 1) allow a person to raise excitement and then reduce it in the form of a reward. The thrill of increasing unpredictability, for example, in a social situation, would be followed by a resolution, a decrease in entropy, which is rewarding. Second, with regard to the relation between salience and reward, one of the ways a stimulus can be salient, after all, is to be rewarding!

In the midbrain we encounter neurons that synthesize and use 5HT. The rate of electrical discharge by midbrain serotonergic neurons classically has been tied to arousal in that firing rates by these cells are correlated with a sleep state.[294] The great French neurobiologist Jouvet, in company with an entire generation of neurophysiologists studying 5HT neurons, showed that the discharges of these neurons would decrease as the animal's state progressed from waking into sleep.[295] Most exquisite have been recordings in freely moving, behaving cats[296] in which a monotonic decrement in firing rate was shown, from active waking through quiet waking through slow wave sleep through rapid-eye-movement (dreaming) sleep. The activities of midbrain serotonergic neurons can be distinguished from those of hindbrain (medullary) serotonergic neurons in that the latter respond to specific motor challenges and to cold temperatures.[297,298] As with DA neurons, feedback plays a role. For serotonin neurons, this means negative feedback by serotonin through 5HT-1a receptors.[299] In summary, from a broad theoretical point of view, Barry Jacobs, whose Princeton lab has done much of the modern work on serotonin neurons in behaving animals, implicates the medullary neurons in coordinating autonomic and neuroendocrine responses to the demand for motor activity.[300] But the midbrain neurons are not closely related to specific motor challenges and instead correlate with the state of the waking/sleeping cycle.[295] Therefore, both serotonergic neuronal groups are positively correlated with behavioral arousal.[301]

Separate from DA and 5HT neurons, in the midbrain reticular formation are nerve cells whose activity is correlated with the arousal state of the animal and is not tied to any particular stimulus modality. In a classical type of experiment, Manohar and Adey found many such neurons that are very active in the aroused, awake animal but that have decreased activity levels during slow wave sleep.[302] Likewise, Kasamatsu[303] recorded from reticular neurons that fired fewer spikes as the animal's arousal level decreased into light sleep, but he also had some cells that showed the opposite trend and were most active during deep sleep. In all of these experiments, the key parameter was not the particular modality of environmental stimuli but the animal's arousal state. For these midbrain reticular formation neurons, stimuli from several sensory sources can converge onto an individual cell. Classical electrical recording studies show widespread receptive fields on the skin as well as multimodal stimulus effectiveness.[304,305]

Following dynamic changes in the lower brainstem systems, certain thalamic neurons that control the state of electrical discharge in the cerebral cortex are also activated. Steriade and his colleagues in Montreal have documented that short bursts of several action potentials are correlated with a relatively inactive cerebral cortex and sleep. In contrast, steady firing of single action potentials triggered by sensory stimuli correlates with transmission of significant stimulus information to the cerebral cortex whose circuitry thus would become activated. There are exceptions in thalamic sensory relay nuclei,[306] but electrical discharges in thalamic reticular neurons, which can gate afferent information headed for the cortex, reflect these two thalamocortical states.[307]

■ Special Cases: Olfaction and Vision

From the neuroanatomy described in Chapter 2, we understand that the olfactory system and the visual system present special cases. Their access to arousal systems must be different from other stimulus modalities.

Olfactory signals to arousal systems are relatively straightforward. As we travel along olfactory axons from main olfactory bulb and pheromone-signaling axons from the accessory olfactory bulb, the first major brain region we hit is the amygdala. Nerve cell groups in the amygdala are intimately involved in the control of behaviors and autonomic functions associated with heightened states of arousal (reviewed, Aggleton[308]; see Chap. 6). Immediately we see how olfactory stimuli and certain pheromones could affect arousal levels associated with, for example, sex and fear. Activation of the human amygdala during sexual excitation[309] and (from the work of Elizabeth Phelps, *Curr Opin Neuro*, 2004) fear proves the point.

For explaining the access of visual signaling to arousal mechanisms, we have two sets of mechanisms. First, consider the deep layers of the optic tectum, also known as the superior colliculi. Neurons here receive a convergence of sensory signals from several stimulus modalities[60] and are clearly related to the initiation of movements, for example, eye movements and reaching movements.[61] Second, the thalamus. Besides the pulvinar nucleus in the thalamus (Chap. 2), consider the midline and intralaminar thalamus and the thalamic reticular nucleus. In the visual sector of the latter cell group are neurons that receive excitatory inputs from primary visual pathway neurons and that are obviously in a position to "gate" visual signals to the cerebral cortex (reviewed, Crabtree[310]). If any neuron type could determine whether a visual stimulus would activate the cortex, this is it. Both of these routes of signaling, the midline and reticular thalamus and the deep superior colliculus, show us how visual signals may not merely trigger specific motor responses restricted to the object perceived but also may change the state of the subject's forebrain.

▒ Informational Content Governs Amplitude of Response in Neurons Related to Arousal

Let's relate all of the electrophysiologic responses recounted in this chapter to my main theme: arousal mechanisms highlight the importance of mathematical calculations of information content. I propose that novel, surprising stimuli produce larger more reliable responses in arousal-related neurons than routine, monotonous stimuli. One way of bringing in unpredictability and uncertainty—to discuss how unpredictability and uncertainty control electrophysiologic responses in certain brainstem neurons—is to understand that some neurons "encode probability" (Glimcher[311], p. 256). We already saw evidence that neurons in arousal pathways respond especially well to salient stimuli. A stimulus that is valued obviously becomes salient. Recording from neurons in the monkey brain, Paul Glimcher and his colleagues at New York University found cortical neurons that responded more strongly to stimuli according to the "expected utility" (p. 264) of those stimuli. "Utility" refers to their value, and "expected" to their probability. In other work with monkey brain, Satoh et al.[51] discovered neuronal responses that were related to reward expectations. Neuronal activities can be understood with respect to the prediction of reward[312] rather than the stimulus itself or the reward itself. "Prediction" is an information concept. Therefore, in relation to arousal pathways, I reformulate standard neurophysiologic ideas to include the dominating influence of information content on response magnitude.

The power of uncertainty, unpredictability, and change to control the sizes of responses to sensory stimuli by neurons is illustrated dramatically by the phenomenon of habituation. Habituation refers to the universal observation that steady repetition of the same stimulus causes a steady reduction of response. Habituation is illustrated by electrophysiologic recordings from individual nerve cells in CNS regions as widely separated as the spinal cord and the hypothalamus (Fig. 3.6). Other experiments show the power of this phenome-

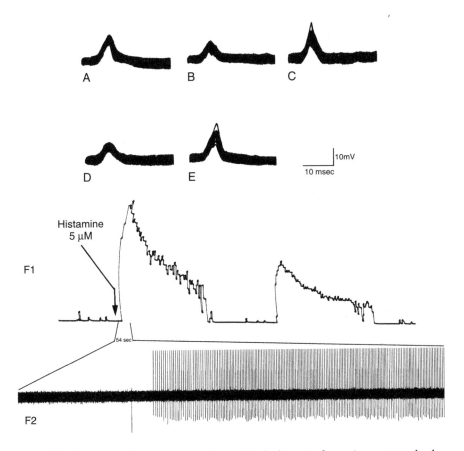

Figure 3.6. Repeated stimulus presentations, with declining information content, lead to decreased response. Similar to the behavioral data in Figure 1.6, we illustrate electrophysiologic demonstrations, from spinal cord motor neuron recording (A–E) to hypothalamic neuron recording (F), of decreased responsiveness upon stimulus repetition.

A–E. Repeated electrical stimuli. The authors recorded an excellent postsynaptic potential from a spinal motorneuron (A). (B) When the electrical stimuli were repeated at a frequency of 1/second, the amplitude of the postsynaptic potential declined. (C) The motorneuron was "woken up" and the response size restored following stimulation of another, quite different nerve. (D) Continued stimulation of the initial nerve at a frequency of 1/sec reduced the amplitude of the postsynaptic potential. (E) The response size was recovered to its initial value by reducing the frequency of the repeated stimulation to 1–3 seconds. (From Spencer, Thompson, and Neilson, *J Neurophysiol*, 1966; 29:253. See also Thompson et al., *Psychol Rev*, 1966; 73:16–43.)

F. Repeated neurochemical stimulus. Recording from a ventromedial hypothalamic neuron, at the top of the lordosis behavior circuit, Lee-Ming Kow saw a very large response to the arousal-related transmitter histamine in the neuron, which had virtually no spontaneous activity. The cell's spike freqency record is on the top line (F1), and a small fragment of the response duration is illustrated with the raw data on the bottom line (on an expanded time scale, F2). Simple repetition of the histamine stimulus (top line) yielded a significantly smaller response. Aside from demonstrating the repetition/response decrement, these recordings also reveal the effect of a generalized arousal transmitter on a neuron connected with a specific form of arousal, sexual arousal. (From Kow and Pfaff, 2004, unpublished data.)

non as it reduces the magnitude of behavioral responses. The very existence of habituation in a wide variety of species proves the power of information theory to explain certain aspects of electrophysiologic signaling.

Information theoretic treatments of neuronal signaling will lead us to a new approach to computational neuroscience. It is already providing useful insights. With respect to GABA neurons in the hippocampus, for example, calculations of variance and entropy provide quantitative parameters that promise new insights into mechanisms of synaptic plasticity and neurologic disease.[313]

■ Cerebral Cortex, the EEG

In the human brain, the inevitable consequence of heightened activity in brainstem arousal pathways is the activation of the cerebral cortex. Since the pioneering work of Donald Lindsley and Horace Magoun at UCLA, we have understood that damage to the ascending reticular arousal systems puts the cortical electroencephalogram (EEG) into a sleep-like state,[23,314] whereas stimulation of these brainstem systems enhances cortical arousal. In humans the cortical EEG is recorded from the surface of the scalp. The electrical waves recorded are thought to result from large groups of synchronous postsynaptic responses in local populations of neurons beneath the scalp electrode.[315] High-frequency low-amplitude activity is associated with behavioral wakefulness and alertness, whereas low-frequency high-amplitude waves begin to be seen as the person goes to sleep. Thus, cortical electrical activity reflects behavioral state transitions.

The distributed cortical systems whose neuroanatomy is described in Chapter 2 generate both the electrical changes and the functional magnetic resonance imaging (fMRI) changes associated with such subtle state changes as "inattention" to "conscious rest."[316] While the temporal patterns of spike discharges in thalamocortical neurons are important for influencing the EEG, neuronal activity in certain basal forebrain cell groups are important as well.[317,318] In our lab these state transitions are recorded in experimental animals, for example, mice, which are tractable to genomic manipulations.

For humans, two of the most important psychological consequences of activation of the cortical EEG are alertness and attention (Posner and Di Girolamo,[319] pp. 623–631). Without cortical electrophysiologic activation, we are not alert and aware of our own acts.[320,321]

■ Electrical Activity in Three Arousal-Related Biological Systems

As a result of electrical activity in arousal systems, fluctuations of activity can be recorded in forebrain neurons related to sleep, sex, and fear. First, for sleep,

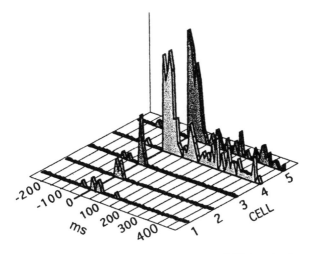

Figure 3.7. The laboratory of Joseph LeDoux at New York University has reported many examples of nerve cells in the lateral nucleus of the amygdala responding to stimuli that signal fear. In these five cells, varying magnitudes and durations of responses are recorded. The neurons had virtually no spontaneous activity. (From Quirk, Repa, and LeDoux, *Neuron*, 1995; 15:1029.)

the neurons in the sleep-promoting ventrolateral preoptic area (vlPOA) in the basal forebrain have been best documented. These neurons' electrical discharges are inhibited both by acetylcholine and by noradrenaline,[322] thus reducing sleep and enhancing wakefulness. Some of the sleep-related physiologic alterations may also be produced by thermoregulatory neurons in the basal forebrain.[323] Their biophysical mechanisms will now come under greater scrutiny, as sleep-promoting neurons can be studied under rigorously controlled *in vitro* conditions.[324]

Second, electrophysiology related to sex has been successfully completed during recordings of neuronal activity in the ventromedial nucleus (VMN) of the hypothamus. This cell group is at the top of the lordosis behavior circuit and controls mating behavior in female quadrupeds.[19] As you might expect, neurophysiologic influences from ascending arousal pathways increase electrical activity in these VMN neurons. Application of norepinephrine (NE) or acetylcholine to these neurons causes them to fire at a higher rate,[92] the former acting through NE α-1b receptors and the latter through muscarinic receptors. An exquisite application of my theoretical approach is to test the relation of histamine (HA) to these VMN neurons because of the strong relation between HA and generalized arousal. Clearly, VMN neurons are excited by HA applications (look ahead to Fig. 6.5). From these experiments, it is easy to see how generalized arousal fosters sexual arousal.

Third, Joseph LeDoux and his colleagues at New York University have clearly drawn electrical activity of neurons in the lateral nucleus of the amygdala (Fig. 3.7) into the mechanisms controlling our responses to stimuli triggering fear.

■

Electrophysiologic recordings provide plenty of evidence for multimodal neurons whose activity corresponds to the behavioral evidence for generalized arousal and can cause changes of state in the CNS. We have seen that these neurons participate in distributed systems, they respond to large sets of high-information stimuli, and they influence entire classes of responses.

4 Autonomic Nervous System Changes Supporting Arousal; the Unity of the Body

Against dogma, dynamic patterns of two autonomic subsystems prepare the body for changes in CNS arousal states. Teamwork, not identity, among autonomic responses fosters information flow and cooperation.

The previous chapter explored how the electrical activation of brainstem and cortical pathways that causes behavioral arousal is produced. Here we delve into parallel changes within the autonomic nervous system—changes in the cardiovascular system and gut, with the corresponding changes in respiration that always accompany emotional behaviors.[325]

These comprise one component of generalized arousal. The complete definition of generalized arousal (Chap. 1) states that its increase entails increased alertness to sensory stimuli, increased voluntary motor activity, *and* increased emotional reactivity. Emotional behaviors depend on certain patterns of autonomic and endocrine adaptations.[326,327]. For example, in a state of anger, blood pressure and heart rate increase markedly; in a state of fear, they increase somewhat; but during depression, they stay the same or decrease. Some of these tendencies can be inherited and their routes of genetic influence traced.[328] In fish,[329,330] mice, rats, dogs,[331,332] and horses, genetic influences on emotional behaviors are obvious. In humans, genetically trained psychiatrists have been able to discern some traits of temperament for which genetic variations play a discernable role.[333–336] What has been called "autonomic arousal" must play an interesting part in our story.

Classically, autonomic nervous systems have been divided into two parts: sympathetic and parasympathetic systems.[337] Sympathetic autonomic responses work through a chain of small ganglia next to the thoracic and lumbar levels of the spinal cord. The neurons of these sympathetic ganglia use norepinephrine as a transmitter and signal through long axons to a variety of organs in the body with an overriding result: the use of metabolic energy to prepare the body for action.[338] The old characterization of this system was "fight or flight." In contrast, parasympathetic autonomic responses work through ganglia more widely distributed away from the spinal column, to be near their target organs. The neurons of these ganglia use acetylcholine as their transmitter, and their major actions tend to have the net result of gaining or preserving metabolic energy.

■ Patterns of Autonomic Responses

Arousal states supporting strong emotions are invariably accompanied by changes in the cardiovascular, respiratory, and thermoregulatory states of our bodies. I argue that the dogma of sympathetic (fight or flight) states always being at war with parasympathetic states must be discarded. A classical example of this dogma came from the examination of the iris in the human eye. Sympathetic nerves cause it to dilate, while parasympathetic nerves cause it to constrict. However, examples of sheer opposition between the two autonomic systems are rare.

I claim that *dynamic patterns* of activity in these two major subdivisions of the autonomic nervous system are essential to drive the expression of emotions such as fear or sexual desire. In Terry Powley's phrase, "the two subdivisions are choreographed" to achieve proper controls over various sphincter muscles in gastrointestinal organs. Likewise, timed cooperation between sympathetic and parasympathetic neurons accounts for penile erection and ejaculation (Fig. 4.1). According to an authority like Kevin McKenna,[339] at Northwestern University, erection has a clear parasympathetic basis, but ejaculation provides an excellent case of cooperation among sympathetic, parasympathetic, and somatic nervous systems. While its contraction of the bladder neck depends

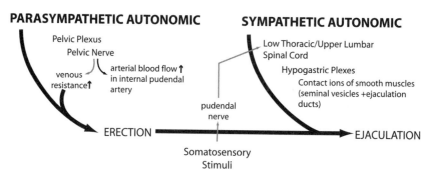

Figure 4.1. Simplified schematic to illustrate how parasympathetic and sympathetic autonomic neuronal stimulation cooperate to produce a biologically important response: penile erection and ejaculation. Adapted, following advice from Kevin McKenna (Northwestern Univ.) from Benson, Chapter 25 (vol. 1, ed. 1, pp. 1121–38); Sachs (2000), and Sachs and Meisel, Chapter 37 (vol. 2, ed. 1, pp. 1393–485), both in *Physiology of Reproduction*, ed. E. Knobil and J. Neill, 1988.

Figure 4.2. The human paraventricular nucleus of the hypothalamus, a nerve cell group important for arousal in the endocrine, cerebral, autonomic, and behavioral domains. Neurons are stained immunocytochemically to demonstrate vasopressin-expressing cells. A, C, E at low magnification. B, D, F at higher magnification. Note the differences in intensity of staining between a young man (A, B), a young woman (C, D) and an old woman (E, F). (From Ishunina and Swaab, *J Clin Endocrinol Metab*, 1999; 84:4637.)

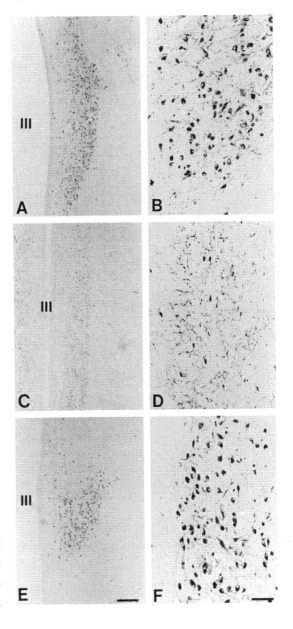

on sympathetic innervation, the external urethral sphincter requires somatic neural stimulation, and the expulsion phase is coordinated by all three systems. In terms of generating patterned autonomic responses, Clifford Saper and his colleagues have noted parallelisms between pain pathways and the visceral sensory pathways that influence autonomic responses,[340] while we have noted (see Fig. 2.2) marked overlaps between pain pathways and sex behavior pathways.[18] We need to find out exactly how these neuroanatomic parallelisms help to organize changes in autonomic states through time.

The hypothalamus orchestrates patterns of autonomic, endocrine and behavioral responses to emotionally significant stimuli.[341,342] In fact, it has been called the "head ganglion of the autonomic nervous system." Within the hypothalamus, an amazing group of nerve cells called the paraventricular nucleus (PVN) of the hypothalamus (in the human, see Fig. 4.2) plays a central role.[343] Originally famous for expressing genes encoding the hormones oxytocin and vasopressin, PVN neurons were later discovered to project to the lower brainstem and to autonomic centers in the spinal cord[344–348] (Fig. 4.3). Thus, the

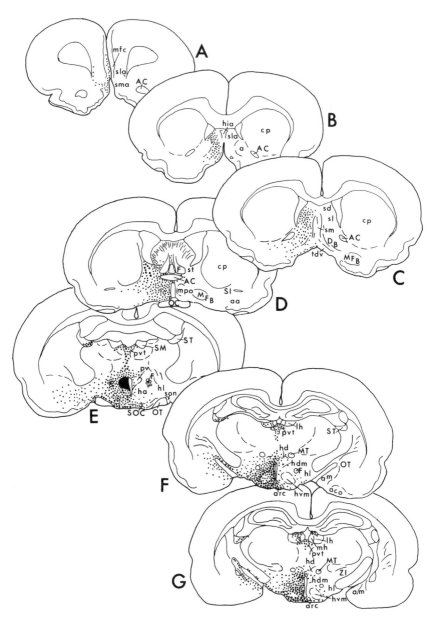

Figure 4.3. Long descending axons from neurons in the paraventricular nucleus (PVN) of the hypothalamus to the hindbrain reticular formation and autonomic control centers. Microinjection of a radioactively labeled amino acid into the PVN (black region, pv) allowed the nerve cells there to make radioactively labeled protein, which was transported by the axons and revealed by autoradioradiography. The cells projected as far posteriorly as we sampled (section N) and included in their distribution the reticular formation (RF) and an autonomic controlling cell group, the nucleus of the tractus solitarius (ns). Clif Saper and Larry Swanson showed similar results and even demonstrated that some PVN axons reach autonomic-related cell groups in the spinal cord. (From Conrad and Pfaff, *J Comp Neurol*, 1976; 169:221.)

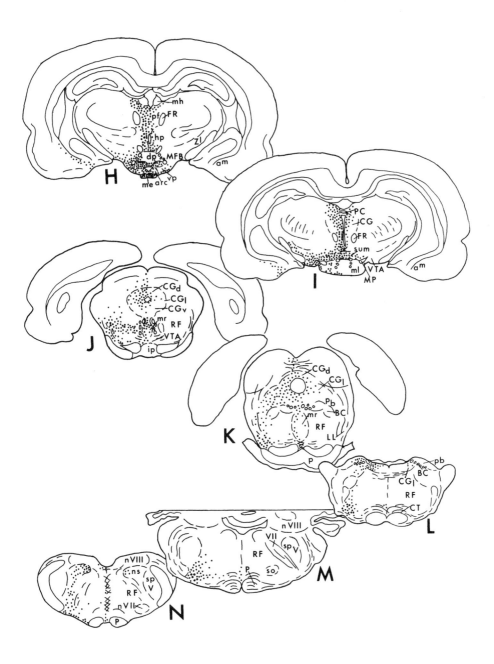

neuroanatomic foundations for hypothalamic coordination of endocrine, autonomic, and behavioral events (even behavioral reward[349]) were laid. Subsequently, PVN neurons have been shown to control critical parameters of cardiovascular physiology[350] through their influences on sympathetic nerve outflow.[351,352] These influences depend on PVN cellular synthesis and release of vasopressin[353–356] and oxytocin.[357] The same holds true for respiration: PVN neurons modulate breathing rhythms by virtue of their peptide products vaso-pressin[358] and oxytocin.[359]

How do PVN's oxytocin- and vasopressin-releasing nerve terminals do their jobs? They modulate a complex network in the lower brainstem that integrates feedback inputs from cardiovascular sensors with visceral sensory inputs and, importantly, with anticipated demand for motor responses (Loewy[360]; see Chaps. 1 and 6, pp. 88, ff). Electrical activity in lower brainstem cell groups such as the nucleus of the tractus solitarius and the dorsomedial nucleus of the vagus determine the responses, for example, to carotid sinus and vagal nerve inputs[361] (see Chap. 10). For blood pressure control,[362,363] some of the master cells in the arousal crescent described in Chapter 2—those in the rostromedial and rostrolateral ventral medulla—are crucial.

Not all hypothalamic controls of autonomic physiology reside exclusively in the PVN. Histamine-expressing neurons exert significant effects over the sympathetic nervous system (e.g., Yasuda et al.[364]). Among other forebrain regions,[365] the amygdala has direct connections to the lower brainstem and clearly manages autonomic responses to emotionally laden stimuli.[366,367]

Some controls over our hearts and blood vessels depend on respiration. Nerve cells in the medulla that have cardiac rhythms of discharge are strongly

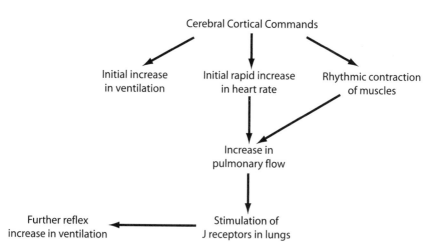

Figure 4.4. A simplified schematic of coordinated changes in respiratory and cardiovascular functions to support vigorous exercise. This scheme emphasizes central CNS commands, but other approaches to the same subject emphasize neurohumoral coordination and reflex controls due to physicochemical parameters controlled by cardiac and respiratory functions. In all theories, autonomic and respiratory responses work together. Adapted from Paintal and Anand, 1998, as presented in Paintal (in *Handbook of Clinical Neurology*, ed. O. Appenzeller, 2000). But also see Wasserman et al., pp. 595–621, in A. P. Fishman, ed. *Handbook of Physiology*, 1986.

modulated by visceral sensory inputs reflecting inspiration and expiration.[368] Human exercise physiology has begun to show how respiratory and cardiovascular adaptations work together to support, for example, rapid locomotion (Fig. 4.4). In different ways, respiratory/cardiac interactions have manifested themselves in all vertebrate species studied.[369] Current studies of respiratory mechanisms suggest that these components of autonomic arousal will soon become clear; for example, we now understand better than ever how brainstem neurons control respiratory rhythmicity.[370] In turn, CNS mechanisms gain additional scientific interest through our knowledge of the physics of breathing, including the dynamics of gas exchange and the mechanisms of rib cage and diaphragm movement.[371,372] My own lab notes with interest the reports that hormones influence the regulation of breathing by both direct and indirect routes.[373] Thus, the respiratory contributions to autonomic arousal offer the opportunity to relate solid physical facts to physiologic mechanisms and to hormonal influences, and finally to correlate with emotional state.

◼ Reformulations

Reformulating theoretical approaches to arousal systems, I give no credit to the old assumption that autonomic arousal uses a completely separate, descending set of mechanisms distinct from those managing arousal of the forebrain. That traditional formulation is neither complete nor correct. Instead, master cells such as those in the arousal crescent of the lower brainstem (and certain neurons in parts of locus coeruleus) contribute to both cerebral and autonomic arousal. These are the large reticular neurons that have strong ascending and descending projections (Chap. 2). Damasio[186] was right. There is a degree of connectivity and coordination heretofore unanticipated.[374] The bipolar nature of autonomic controls in the brainstem—extensive ascending and descending connections orchestrating the system adequately[187,375,376]—encouraged me to formulate the BBURP theory in Chapter 7, as one of the Bs stands for "bipolar" or "bidirectional."

This new theory of autonomic arousal embraces two ideas that could be seen as contradicting each other, but do not. One idea emphasizes the teamwork between autonomic preparations for emotional behaviors, cerebral cortical activation due to emotionally laden events, and the behaviors themselves. The other idea reveals the degree of uncertainty—maximization of information flow—surrounding emotion and arousal.

First, teamwork is required for cardiovascular and respiratory systems to support the muscular requirements for the expression of emotional behaviors (Fig. 4.4). Thus, my formulation is as follows: autonomic, cortical, and behavioral arousal must be coordinated, but not correlated at any given time. This is teamwork, not identity.

Second, emotions and arousal are products of sudden and unpredictable changes. In the terms of this book, they respond to a *high-information environment*. For example, gambling, an activity that embodies human responses to uncertainty and unpredictability, causes autonomic arousal.[377] And the responses themselves have high-information content because they are not perfectly correlated at any moment. In fact, the very lack of correlation—for example, between EEG changes and electrocardiogram changes—helps to guarantee that at least one portion of the body's arousal systems will be sensitive enough to respond to emotionally important environmental changes. We are dealing with arousal systems that employ both ascending and descending signaling components (Chap. 7). To ensure that a person reacts to important, salient changes in the environment, we have two options: (1) A hypersensitive *ascending* arousal system brings autonomic systems into play indirectly, or (2) *descending* arousal systems lead the way—"the gut reacts first"—and autonomic arousal changes eventually force the attention of cortical arousal systems. Either way, the person begins to do what is required.

■ A High Information System Shows Coordination sans Correlation

The classical question of how we accomplish difficult tasks and mental work remains an interesting one.[378] Arousal, alertness, and attention are certainly part of the answer. If you remember the problem attacked and solved in Chapter 1, questions about arousal mechanisms had suffered from a lack of definition and a false dichotomy—"all arousal mechanisms are part of the same monolithic whole" versus "there is no such thing as arousal." Now that we have a clear operational definition of generalized arousal (Chap. 1), the question of its internal structure can be asked anew. The answer is much more subtle than the older dichotomous point of view would have anticipated.

On the one hand, autonomic components of arousal systems must work together with other components in order to provide vascular, respiratory, and metabolic support for effortful behavioral activities. An old literature shows that modification of behavioral activities correspondingly modifies cardiac output.[379] Anticipation of increased or decreased motor output increases or decreases heart rate, respectively.[380,381] Coupled with experimentally produced anxiety, respiration and heart rate decrease.[382-384] Only the human subjects who showed significant changes in muscular activity experienced the heart rate changes.[385] During experiments that examined subjects in greater physiologic detail, cerebral cortical recordings correlated best with heart rate precisely during those time periods when heart output was accelerating or decelerating.[386] In a more complex experiment, Webb and Obrist[387] showed mutual correlations among electromyograph recordings, heart rate, eye blinks, and gross movement during a task that required arousal and attention.[388] Recently, a striking coordination between locomotor and respiratory rhythms has been reported and its basis in brainstem physiology discussed.[389,390] All of these experimental protocols have documented that behavioral arousal and autonomic arousal have a lawful relation to each other.

On the other hand, one of the early researchers in this field, John Lacey, at the University of Wisconsin, reported experimental results in which different aspects of autonomic and behavioral arousal responses were *not* correlated with each other.[391] While he admitted that, in general, autonomic, EEG, and behavioral arousal "occur simultaneously," he claimed that their mechanisms could be dissociated by selective CNS destruction and pharmacologic manipulations. Further, during experiments requiring behavioral activation by virtue of a visual attention task, correlations among different measures of autonomic arousal were low and variable.[392] Likewise, Tursky et al.[393] reported that two measures of autonomic arousal, skin resistance and heart rate, did not show the same time course of response to an experimental behavioral task. Their conclusion from this body of work was that autonomic arousal patterns depended on previous and subsequent overt behavior. In addition, we know that the time constants of different autonomic mechanisms differ. To the extent that autonomic responses to external situations are uncorrelated, there are two implications: (1) an intrinsically high information content in this aspect of arousal neurophysiology; and (2) an adaptive time-sharing of maximal sensitivity that protects against autonomic sluggishness in time of danger.

What we have, therefore, are autonomic functions that are physiologically coordinated to "do the job," but that are not correlated with each other at time *t*. By analogy, consider a football team. Even though the players are following agreed-upon plays and are moving effectively with respect to each other, attempts at correlating their responses statistically at any given time would yield no correlations. Their movements are cooperative, but not identical (Table 4.1). The same could be said for autonomic responses, especially with respect to other components of arousal. Sensory inputs can trigger "teamed," patterned responses, arranged along a timeline. Because the baseline states of the autonomic components are not necessarily equated and because of the low correlations among exact timings of responses, the system essentially embodies high-information content.

Another index of the information-related features of autonomic nervous

Table 4.1. Theoretical relations among autonomic and other arousal responses

	Identical regulation of all measures	Autonomic	"Chaos"
Simultaneity	Yes	No	No
Correlation	Perfect	Low	Zero
Coordination	High	High	Zero
Flexibility	Low	High	Zero
Effectiveness	Low	High	Zero
Information content	Zero	High	Maximum

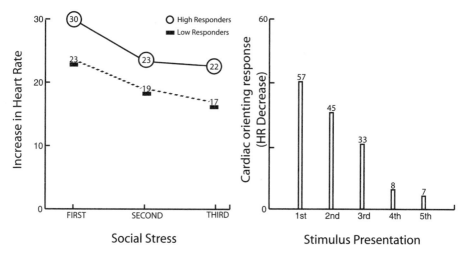

Figure 4.5. Illustrations of habituation in autonomic responses, under two different experimental conditions. These show that information content in the stimulus influences autonomic arousal. *Left*: In men and women, repetition of social stress led to a decline of the increase in heart rate. High responders' average results are shown in circles (30, 23, 22). Low responders' average results are shown underlined (23, 19, 17). Adapted from Schommer et al., *Psychosom Med*, 2003; 65(3):450–60. *Right*: The cardiac orienting response in neonatal rats decreased upon repetition of an olfactory stimulus. Adapted from Hunt and Phillips, *Alcohol Clin Exp Res*, 2004; 28(1):123–30.

responses is their tendency to decline in the face of a repeated stimulus whose informational content, by definition, has decreased. Autonomic responses habituate (Fig. 4.5). Here are some examples: In adult men and women, on repetition of a social stress, there is less of an increase in heart rate.[394] In neonatal rats, the cardiac orienting response waned on repetition of an olfactory stimulus.[395] Mauss et al.[396] recorded self-reported anxiety as well as blood pressure,

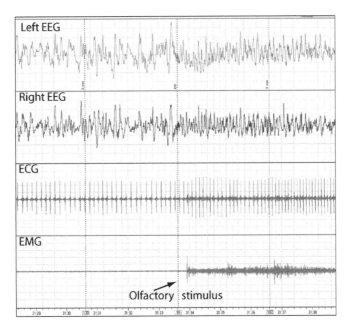

Figure 4.6. Simultaneous electrophysiologic recordings of electroencephalogram (EEG), electrocardiogram (ECG), and neck muscle electromyogram (EMG) in a mouse. Experiments were conducted in the mouse's home cage during the light part of the daily cycle, so that baseline activity would be steady and quiet. Upon presentation of a novel olfactory stimulus, the dominant EEG state changed from high amplitude/low frequency waves to low amplitude/high frequency. The heart rate (ECG) sped up, and the mouse contracted its neck muscles (EMG). This type of experiment opens up arousal neurophysiology in a genetically tractable mammal. (From Stavarache and Pfaff, unpublished data, 2004.)

heart rate, and skin conductance in adults who had been rated as highly socially anxious or not anxious. Despite differences in self-reported anxiety, the habituation and recovery of the autonomic measures were comparable between the two groups.

■ Supporting Hormone-Dependent Behaviors

Hormones affect autonomic physiology in a way that supports hormone-dependent behaviors. Sex hormones are accumulated and retained by neurons of autonomic ganglia[397–399] as well as by cells in the heart and blood vessels.[400–402] The roles of sex hormones leading to sexual arousal (Chap. 6), erections, and ejaculation are patently obvious (see Fig. 4.1). Sex hormones alter cardiovascular responses to activation of medullary neurons in the arousal crescent[403] and address oxytocin neurons in the PVN of the hypothalamus, which project back to the spinal cord and support erection.[404,405] Inhibiting sympathetic outflow in female rats decreases their sex behavior.[406]

We have new data about effects of hormones and hormone receptor genes on autonomic function, measured electrophysiologically (Fig. 4.6). So far, during recordings from mice, it appears that different measures of autonomic versus EEG arousal embody high-information content due to their relative lack of intercorrelation. Correspondingly, according to these data, one arousal response system—for example, autonomic or cortical—is maximally sensitive at any given time and thus guarantees a prompt response to emotionally disturbing stimuli.

■ Summary

We have autonomic nervous systems whose activities accompany emotional arousal and that react to uncertainty, surprise, and change. They manage gross changes of state of the entire body, coordinating visceral, vascular, and metabolic responses with the state of the CNS. Nerve cells in the hypothalamus produce *patterns* of autonomic responses: Sympathetic and parasympathetic autonomic subsystems do not always oppose each other but instead display exquisitely timed cooperation. Information theory helps us to understand how they work. Relations among different autonomic measures embody high-information content, and autonomic responses are maximized in a high-information environment.

5 Genes Whose Neurochemical Products Support Arousal

Recent molecular studies reveal a plethora of genes that regulate arousal states. Some are associated with the classical systems of Chapter 2. Others are brand new, including hormone-influenced genes and a gene whose mutation causes narcolepsy.

For working out the functional genomics of the CNS, behavioral measures offer extraordinarily sensitive indicators of genetic change. Behavior constitutes the bottom line, the most important consequence of CNS gene expression. Among all the behaviors open to investigation, I argued in Chapter 1 that the most fundamental and crucial are those connected to arousal.

All of the neurochemistry of Chapter 2 depends on the differential expression of genes coding for the enzymes that synthesize specific neurotransmitter and neuropeptide ligands, for their receptors, and for their transporters that suck them back out of the synaptic cleft, as well as for the enzymes that chemically break down the neurotransmitters. We already understand some things about a tremendous number of these gene products. The field is exploding with opportunities for new molecular neurobiological work. In this chapter, rather than providing a simple catalog of genes and their products, I combine brief descriptions of classical and novel genes with the concepts that organize their functions. The very large number of genes covered—more than the 124—proves, once again and in a different way, the importance of the arousal functions defined in this book. Through these disparate and overlapping genetic mechanisms, Nature makes absolutely sure that the transcriptional activities required for arousal will not be found wanting.

Genes Associated with Classical Systems

In this section, we highlight some of the main features of those genes whose products increase arousal and then of those genes whose products decrease arousal.

These Gene Products Serve in Systems that Heighten Arousal

Norepinephrine (NE) and Noradrenaline (NA). Tyrosine hydroxylase and dopamine-β-hydroxylase are synthetic enzymes necessary for the synthesis of NE from tyrosine. When that has been accomplished and NE has been released from an axon terminal, there are several types of receptors that can help transduce the signal into a postsynaptic neuron. Receptor subtypes all representing separate gene products include three α-1, two α-2, and three β-adrenergic receptors. They all signal through G-proteins, including the Gq/11 protein, but have different localizations within the postsynaptic cell and different signal transduction preferences.[407,408] Their polymorphisms are associated with phenotypes of clinical significance, including those related to arousal: hypertension, mood disorders, metabolic changes, and sensitivity to adrenergic agonists.[409] Protein kinases not only play an important part in the amplification of the NE receptor signal but also desensitize the receptor after the signal has passed.[410]

From the point of view of arousal neurobiology, it is fascinating that both the synthetic enzymes and receptor sensitivities in the brain can be altered by sex hormones and by fear.[208,411–413] Polymorphisms of adrenergic receptor genes could contribute to temperamental differences in humans.[409,414] It follows that one of the targets of the NE-induced cellular signaling—cyclic AMP response element–binding protein (CREB)—should be involved in arousal of the cerebral cortex, and it actually is.[415]

Finally, genes for metabolizing NE—monoamine oxidase B; catechol O-methyltransferase (COMT)—and that for removing NE from the synaptic cleft to limit its action—norepinephrine transporter (NET)—provide additional opportunities for the regulation of arousal in animal and human brains.[416–418]

Dopamine (DA). The gene that codes for tyrosine hydroxylase produces the synthetic enzyme that is rate-limiting in making DA. There are five DA receptor genes. DRD1 signals through G-proteins, which lead to the activation of protein kinase A. In contrast, DRD2 can have inhibitory effects when it acts presynaptically as an autoreceptor. Its 3D structure has been reported,[419] and it has been implicated in a variety of addictive disorders.[420] The frequent biochemical result showing DRD1 effects opposing DRD2 supports the idea that gene duplication products[421] and differential mRNA splice products can have opposite cellular effects.

DA actions are limited by re-uptake by DA transporters[422] and by chemical breakdown. DA transporters are transmembrane proteins, with cytosolic N- and C-terminal regions separated by twelve membrane-traversing domains, which can be expressed both in nerve cells and in glial cells.[417] They limit the intensity of DA signaling by removing DA from the synaptic cleft. Their sensitivity to psychostimulants, among other characteristics, has made them a popular subject of molecular[423] and neurochemical[424] investigations. Null deletions of a dopamine transporter produce mice that are hyperactive, confirming the expected role of DA in this aspect of arousal (reviewed, Tan, Hermann, and Borrelli[425]). Their polymorphisms could help to account for individual differences in mood and temperament.[423] Two gene products are primarily responsible for metabolizing DA: catechol O-methyltransferase (COMT) and a monoamineoxidase (MAO) type B.

Serotonin, 5HT. Synthesizing serotonin from the dietary amino acid tryptophan requires the gene for tryptophan hydroxylase (TPH). Once 5HT is released from a synapse, no fewer than fourteen gene products coding for serotonin receptors are available to transduce different kinds of biophysical effects (Table 5.1). How they contribute in their separate ways is not yet understood. Serotonin can be broken down chemically by the enzyme MAO type A. However, a compound of much greater popular interest is the serotonin transporter, which ships the transmitter back across the membrane, thereby removing it from the synaptic cleft and limiting its activity. Serotonin's popularity derives from the antidepressant class of selective serotonin reuptake inhibitors (SSRIs), which work by blocking its action. Some studies have claimed an association, in humans, between the serotonin transporter genomic structure and susceptibility to anxiety or depression.[426–428]

Serotonergic systems present clear targets through which hormones can change arousal and mood. Not only do estrogens, for example, increase TPH gene expression and protein level in the monkey midbrain, as shown by

---

Table 5.1. Examples of diverse functions among serotonin receptor gene products

Ligand-Gated Ion Channel Receptors
    Example: 5-HT$_{3A}$ receptor $\Rightarrow$ Na$^+$ and Ca^{++} in $\Rightarrow$ rapid depolarization
G-Protein Coupled Receptors
Signaling through cyclic AMP
    Example: 5-HT$_7$ receptor coupled to G$_{i/o}$, Adenyl. Cyclase $\Downarrow$. Modulate fast psps.
    Example: 5-HT$_7$ receptor coupled to G$_s$, Adenyl. Cyclase $\Uparrow$. Modulate GABA actions.
Signaling through phospholipase C
    Example: 5-HT$_{2B}$ receptor coupled to G$_q$, PLC $\Uparrow$. Nitric oxide release.

*Reference: Hoyer, Hannon, and Martin, *Pharm Biochem Behav*, 2002; 71: 533.

Cynthia Bethea and her colleagues in Oregon, but also they decrease opportunities for serotonin removal. They ramp down mRNA levels for the serotonin transporter and for MAO. Further, George Fink, in Edinburgh, discovered that estrogens increase gene expression for certain serotonin receptors. Thus, the genomic basis for sex hormone effects on mood has been laid. In humans, polymorphisms in the genes for serotonin receptors lead to ideas of how individual differences in serotonin signaling could contribute to differences in temperament.[429]

Acetylcholine (ACh). ACh, whose synthesis is catalyzed by the gene product for the enzyme choline acetyltransferase, helps to activate the cerebral cortex from at least two subcortical sites. From the lateral dorsal nucleus of the tegmentum (LDT), its projections to the thalamus help regulate the sleep-wake cycle. And ACh neurons account for more than 50% of basal forebrain neurons that activate cortical activity, according to the histochemical results of Laszlo Zaborszky and his colleagues at Rutgers Medical School. Genes most obviously involved in the production of cortical ACh receptors are called muscarinic receptors. There are five, with the M1 subtype having an especially strong role in the cerebral cortex. However, ACh receptors with a different set of ligands, or nicotinic receptors, also play a role. ACh is broken down by the enzyme cholinesterase, thus limiting the duration of its synaptic action.

Histamine (HA). Histamine clearly maintains behavioral and cortical EEG arousal.[430,431] From its site of synthesis in tuberomammillary neurons deep in the hypothalamus,[432-434] HA controls wakefulness through widespread ascending and descending axonal projections, but especially through those to the ventrolateral preoptic area.[176] It synthesis requires the gene coding for the enzyme histidine decarboxylase (HDC). Knocking out this gene yields an animal with decreased arousal—behaviorally and in terms of cortical EEG—as well as an increased tendency to sleep.[435]

HA receptors are produced by at least three genes. The H1 receptor gene products strongly depolarize neurons by increasing currents in certain sodium channels and in certain calcium-activated channels, by decreasing "leak" currents in potassium channels, and by other mechanisms as well.[431] Which current is most important for a histamine effect through H1 receptors depends on the neuron type being recorded. HA 1 receptor knockouts exhibit decreased activation of behavior.[436] Reflect, for a moment, on the effects of antihistamines—they are H1 receptor blockers that make us sleepy (Fig. 5.1). While the H2 receptor's activation can also lead to neuronal excitation, like H1,[431] there are physiologic circumstances in which H2s lead to decreased firing rates

EFFECT OF AN H1 RECEPTOR ANTAGONIST
ON SENSORY RESPONSIVENESS

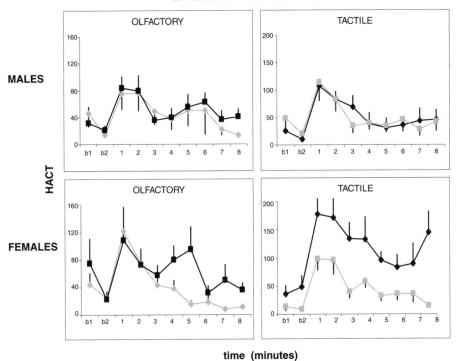

Figure 5.1. An H1 receptor blocker can reduce behavioral responsivity of female mice to external stimuli, as measured by an efficient assay of generalized arousal conducted in the animal's home cage during the animal's daily period of quiescence.[164] Black line = vehicle injection. Grey line = H1 receptor blocker injection. b1 and b2 measure the last two minutes of baseline activity (horizontal activity, HACT) before stimulus presentation. Then, in this example, an olfactory or a tactile stimulus was presented. Where the H1 blocker had no effect, note the quantitative precision of the assay. Also note that females' arousal responses were more sensitive to H1 receptor antagonism than males', a likely effect of genetic cascades starting from the Y chromosome. (From Easton and Pfaff, *Pharmacology, Biochemistry and Behavior*, 2005, in press.)

and, in fact, they can oppose H1's effect. The H3 gene codes for an autorecep-tor regulating HA release. Finally, the metabolism of HA depends on the gene for histamine N-methyltransferase (HNMT) followed by MAO-B.[437]

Systems that wake up the brain sometimes act through other neurochemi-cal pathways for the rapid activation of behavior. Orexin/hypocretin, a newly discovered gene involved in sleep-wake control, probably derives part of its neurophysiologic power through orexin-2 receptors[438] exciting HA neurons in the hypothalamus.[439]

Glutamate receptors. Master cells, which are central to brainstem arousal mechanisms (Chap. 2), use the amino acid glutamate as their transmitter. Genes coding for glutamate receptors come in two flavors: ionotropic and metabotropic. There are at least six gene families comprising at least sixteen genes that code for ionotropic glutamate receptors.[440] Alternative splicing and editing of the respective RNAs add still more functional ionotropic receptors.[441] They are divided into three groups called AMPA, KA, and NMDA receptors according to their affinities for various glutamate-like ligands. Different gene products, even within the NMDA category, have different effects on synaptic plasticity.[442] Disorders of ionotropic glutamate receptors may contribute to some forms of chronic pain[443] and even schizophrenia.[444] Opening ionotropic glutamate receptors excites cells by allowing the positively charged ions sodium and potassium into cells, thus depolarizing them.

In contrast, metabotropic glutamate receptors, binding their ligand in the manner of a Venus flytrap,[445] form three families composed of eight genes.[446] One of these families signals by coupling to Gq/11 proteins, which act through inositol phosphate. The other two families are Gi or Go coupled, and a prominent result is the decrease of activity of the enzyme adenylate cyclase. These actions result in synaptic modulation and slow excitatory or inhibitory synaptic potentials.[447]

Glutamate action is limited, in part, by five glutamate transporters that have various distributions among CNS neurons[448] as well as by vesicular glutamate transporters for which there are two genes. The amino acid sequences of these gene products have diverged enormously from their original family of genes, but a 150–amino acid sequence in the C-terminal part of these proteins has been somewhat conserved.[449–451] These are responsible both for recapturing endogenously synthesized transmitter and for taking up transmitter released from neighboring cells.

These Gene Products Serve in Systems that Reduce Arousal

Adenosine. Adenosine receptors are prominent in daily life in that caffeine blocks them, thus keeping us awake and alert. Adenosine is produced from the common intracellular signaling molecule adenosine monophosphate (AMP) by the enzyme 5'-nucleotidase.[452] It works through any of four receptors, all of which have been cloned and are expressed in the brain.[453] They are all glycoproteins. In some cases their physiologic action draws them directly into the explanation of arousal. For example, the expression of the gene for the A1 receptor in the lateral dorsal nucleus of the tegmentum (LDT, an important cholinergic cell group; see Chap. 2) sufficiently accounts for our tendency toward sleep near the end of our active daily hours. Activated adenosine receptors work selectively through G-proteins, with A1 receptors coupled to Gi and A2 receptors to Gs.[453] Eventually, their physiologic actions modify both potassium and calcium channels.

A striking action of the A1 receptor inhibits the release of classical neurotransmitters, for example, by reducing certain calcium currents in cholinergic neurons.[454] The A2a receptor gene is expressed primarily in regions of the brain innervated by dopamine. One of its intracellular effects is to stimulate the phosphorylation of a phosphoprotein abbreviated DARPP-32, a nodal point in cellular signaling affected by dopamine, serotonin, and glutamate.[455] In the preoptic area of the basal forebrain, activation of A2a receptors represents a route of action of prostaglandin D and puts the animal to sleep.[456]

Opioids and their receptors. Three genes code for major opioid neurochemical systems: The gene for pro-opiomelanocortin yields a small protein, which, when cleaved, produces β-endorphin. The other two genes make the mRNAs for the enkephalins and for dynorphin. What about opioid receptors? The gene for the μ-opioid receptor makes the only protein through which β-endorphin acts effectively but also can help signaling by the enkephalins. The gene for the δ-opioid receptor produces a compound that primarily transduces effects of the enkephalins but also works for dynorphin. The gene for the κ-opioid receptor is quite devoted to dynorphin as a ligand. These opioid receptors can signal through G-proteins in a somewhat promiscuous manner, eventually inhibiting neuronal activity. While the most obvious biological use of these opioid peptides is to reduce pain,[18] they also reduce arousal.

The significant similarities between the neural circuits controlling the sex behavior, lordosis, and those reducing pain[18] tell us how specific aroused states related to sex and pain could interact with each other. According to Garey and colleagues,[26] elevations of these specific emotional states increase overall arousal. We return to this idea in Chapter 6.

γ-aminobutyric acid (GABA). Perhaps the most complex and diverse genetic contributions to the regulation of arousal come from the neurotransmitter GABA. It is synthesized from glutamic acid through the catalyzing effects of two gene products coding for glutamic acid decarboxylase, GAD 65 and GAD 67. Working through the more obvious of the two sets of mechanisms, the transmitter clearly reduces arousal. At least sixteen different gene products[457] produce the α, β, and γ subunit proteins that make up the GABA-A receptor.

They clearly contribute to the regulation of the human electroencephalogram (EEG).[458] Many anesthetics use the GABA-A receptor.[459] For example, ion channel gating properties of inhaled anesthetics seem to require a normal α-1 subunit gene product.[460] Benzodiazepines relax a person by potentiating the ability of GABA to impact the GABA-A receptor in order to increase chloride conductance across the nerve cell membrane. Their effectiveness in doing so depends on a specific gene product: that for the α-2 GABA-A receptor subunit.[461] Alcohol has its soporific effects by virtue of the GABA-A receptor as well. In fact, a specific group of GABA neurons, those in the ventrolateral preoptic area of the basal forebrain, are required for us to sleep.[166,174,176]

But GABA-related gene products also can work in the opposite direction on arousal by GABA neurons inhibiting other GABA neurons—a process called disinhibition. Steve Henriksen and his colleagues at the Scripps Research Institute have discovered GABA neurons in the ventral tegmental area that have long axons projecting to the basal forebrain that could disinhibit the GABA neurons in the ventrolateral preoptic area,[462] thereby awakening the animal. Laszlo Zaborszky reports that a full 25% of basal forebrain neurons responsible for "waking up" the cerebral cortex are GABA neurons that work through other GABA neurons in the cortex.[84] Thus, the eighteen-plus genes responsible for normal GABA signaling can regulate arousal in either direction, depending on how their neurons are hooked up.

■ Genes Newly Recognized

Other genes involved have either been cloned only recently or newly recognized as participating in CNS arousal mechanisms.

Orexin/hypocretin. Discovered as the genetic modules whose mutations cause narcolepsy, orexin/hypocretin systems are receiving much attention from neuroscientists. Coded by two genes cloned and sequenced in the laboratories of Masashi Yanagisawa in Dallas and Tsukuba, and Greg Sutcliffe at the Scripps Institute, orexins strongly enhance wakefulness.[463,464] Gene structures of two exons and one intron, comprising about 1400 base pairs, are well conserved among species. The two peptides consist of thirty-three amino acids (orexin A) and twenty-eight amino acids (orexin B). Null mutations of these genes disrupt wake/sleep rhythms, as demonstrated so far in mice, dogs, and humans.[465] Conversely, orexin peptide administration "rescues" the narcolepsy phenotype of orexin null mutant mice.[466] Expressed in lateral and perifornical areas of the hypothalamus,[467,468] their receptors are widely distributed in the CNS. Two receptor genes have been found, coding for proteins that are structurally similar to G-protein-coupled neuropeptide receptors.[469] Mutations of the Hcrtr2 receptor gene in dogs causes narcolepsy.

Part of the initial excitement at the cloning of the orexin/hypocretin genes derived from their actions on feeding. Food deprivation activates orexin/hycretin neurons.[470] In turn, microinjection of orexin A into the lateral hypothalamus stimulates feeding.[471] I conclude that orexin gene products help to regulate arousal appropriate to the hunger state of the animal.[472] They illustrate how a specific arousal state can influence generalized arousal.

We see how orexins can rapidly change the state of the CNS by considering their multiple projections—"feed-forward" systems—to other arousal-related CNS systems.[473] One important route of orexin action works through projections to an important arousal-transmitter, histamine.[183,474] But the histamine connections are only a small part of the story. Orexins also activate CRH neurons[475] (see the following information on CRH), serotonin neurons,[135] paraventricular hypothalamic neurons,[476] basal forebrain cholinergic neurons,[477] and, importantly, neurons of that rich source of noradrenergic systems, the locus coeruleus.[282,478,479] We also see the evidence for positive feed-forward actions in the effects of orexins on neuroendocrine stress systems[480] and autonomic control systems.[481,482] Putting all of this knowledge about orexin neurobiology and genetics together, we understand how orexins activate many other arousal-producing neurons, thus heightening the arousal state of the entire CNS, especially when the animal does not have enough food.

Prostaglandin D. Prostaglandin D is synthesized from prostaglandin H by two synthases, one the so-called lipocalin-type prostaglandin-D synthase (PGDS). This latter transcript is induced by estrogens in the hypothalamus and depressed in the preoptic area,[483] where it reduces arousal[484–486] (Fig. 5.2). Prostaglandin D's actions eventually require routes through adenosine systems, specifically adenosine 2A receptors, as its sleep effect[487] is blocked by an antagonist of that receptor subtype but not, for example, adenosine 1 receptors.[488] Moreover, these are selective actions of prostaglandin D, because prostaglandin E activates HA systems to awaken the animal.[489]

I distinguish sleep induction by prostaglandin D from the daily effects of circadian clock genes. Transcriptional feedback loops controlling daily rhythms have been worked out (Fig. 5.3) in fruit flies and mice.[184,185,490] The mammalian circadian clock controlled by neurons in the suprachiasmatic nucleus of the hypothalamus acts as a gate that changes the threshold for the actions of the arousal-producing and arousal-reducing gene products we are dis-

Blotted total RNA was probed with a P^{32}– labeled DNA probe for PGDS.

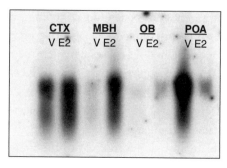

PGD$_2$S

The same blot was reprobed with a DNA probe for 18s ribosomal RNA as a measure of RNA quantity per lane.

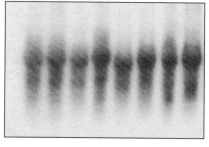

18s

Figure 5.2. Jessica Mong, using Affymetrix microarrays, discovered an estrogen regulated mRNA, that coding for lipocalin type prostaglandin-D synthase (PGDS). We were intrigued by the opposite regulations comparing two nearby brain structures: estrogen-caused increases in the medial basal hypothalamus (MBH), but a significant decrease in the preoptic area (POA). Interfering with PGDS transcript function by locked nucleic acid antisense oligomers microinjected directly into the POA, as predicted, elevated arousal-related motor activity and increased lordosis behavior. (From Mong et al., *Proc Natl Acad Sci USA*, 2003; 100(1):318–23 and 100(25):15206–11.)

cussing in this chapter (look forward to Fig. 7.3). In fact, homeostatic sleep regulation is preserved in mice whose important clock genes *period 1 and period 2* have been deleted.[491] Thus, circadian clock genes provide relatively invariant gating functions, while the gene for PGDS is related to sleep drive, a highly regulatable function.

CRH. The gene that produces corticotropin-releasing hormone (CRH) is expressed strongly in the paraventricular nucleus of the hypothalamus and certain other parts of the basal forebrain.[492,493] Clearly involved in mounting responses to stress by the brain—through both endocrine and behavioral routes—CRH elevates arousal.[494–496] Two genes code for its receptors, CRH-

R1 and CRH-R2, the former a rapidly acting system whose responses are necessary for high anxiety. Stress-induced behaviors require CRH receptors.[497]

CRH is a perfect example of a gene that has evolved to do something else but whose product bears on CNS arousal. Another example is arginine vasopressin (AVP). Vasopressin, a nine–amino acid neuropeptide, helps to alert both the forebrain and the autonomic nervous system when the animal's or human's salt and water balances are challenged. When extracellular salt concentrations are too high,[498–500] the vasopressin gene is turned on. Its peptide product works in the kidney through V2 receptors to help retain water and, in the CNS, primarily through V1a receptors to affect behavior. AVP shows us how causing a specific source of arousal, thirst, can heighten generalized arousal.

Nuclear receptors. Genes for estrogen receptors clearly serve reproduction and behaviors associated with reproduction,[20] but one of them also has significant effects on generalized arousal. Null mutations in the classical estrogen receptor-α reduce an animal's response to sensory stimuli (see Fig. 1.3) and voluntary motor activity (see Fig. 1.4). The same can be said for thyroid hormone receptors, which are fundamental regulators of the body's metabolism. Hyperthyroid patients often become agitated, irritable, restless, and insomniac, while hypothyroid patients are usually sluggish and depressed.[501,502]

If there is any system in the body as complicated as the nervous system, it is the immune system. During infections, the increased desire to sleep has been widely recognized and depends in part on interleukin gene products such as IL-1,[503] IL-6,[504] and IL-8.[505] IL-1 acting in the amygdala leads to behavioral depression.[506] Biochemical routes of action include CRH[507,508] and an alteration of tryptophan metabolism that affects 5HT signaling. Blocking serotonin-2 receptors reduces IL-1 effects on sleep.[509] In fact, the two-way actions between IL-1 and CRH can produce fast changes of state.[510] Both IL-1 receptor gene products and IL-1 receptor–associated proteins are required. I expect even more complexity from this set of molecular mechanisms for arousal, as there are more than forty cytokines and at least fourteen serotonin receptor genes (reviewed, Krueger and Majde[511]).

When I consider the array of genetic systems contributing to the normal regulation of arousal—immune system receptors, genes related to water balance, and so on—I am reminded of the phrase of the classical physiologist Walter Cannon: "the unity of body." When medical doctors such as Dean Ornish or Andrew Weil talk about mind-body connections, would the connections between immunologic signals and arousal systems be an example of what they mean?

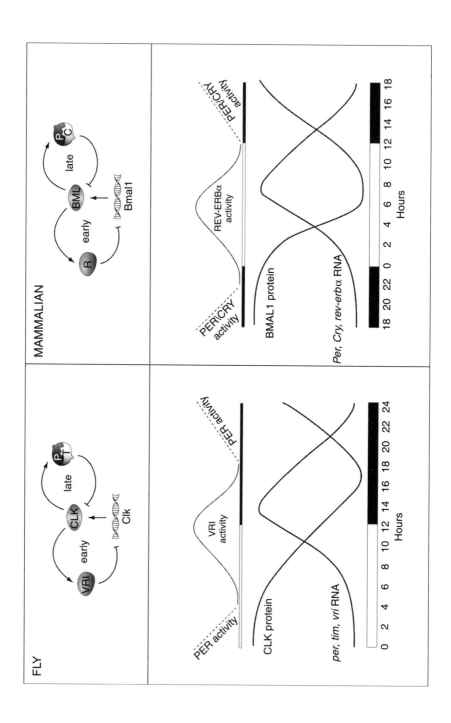

■ Concepts and Questions

New genome-wide scanning techniques—microarrays,[512] differential display,[513] suppressive subtractive hybridization, and quantitative trait loci[514-517]—are continually being applied to finding genes for behaviors related to arousal. What are the benefits of these broad-scale efforts that begin with no neurobiologic ideas in mind? Clinically, these efforts offer the possibility of brand new ideas about etiology and treatment of the many conditions involving disorders of arousal (see Chap. 8). Such ideas—often unimaginable at first, from traditional neurophysiologic points of view—might also help to predict which patients are at risk for such disorders, understand interactions with certain environments, and spell out mechanistic routes of new therapies. In addition to these potential medical benefits, at least four intellectual lessons can immediately be drawn.

1. Multiplicity proving essentiality. The number of genes coding for synthetic enzymes for transmitters and peptides related to arousal, for their receptors, transporters, and catabolic enzymes (reviewed here, at least 124 genes), reminds us that this elementary neurobiologic function is extremely important. Nature will not allow arousal regulation to go wrong. Besides sheer multiplic-

Figure 5.3. Transcriptional feedback loops—incorporating negative feedback with a time delay—at the basis of circadian rhythms in flies (*Drosophila*, left) and mammals (mouse, right). This simplified diagram reveals similarities between widely divergent types of animals. Mechanisms beyond transcription include protein phosphorylation and regulated entry of proteins into the cell nucleus.

Fly clock (*left*). *Top:* Negative feedback cycle in transcription. P = *period* gene. T = *timeless* gene. *Middle:* High levels of VR1 activity block the expression of genes for period (PER) as well as *timeless*. Entrainment by the light cycle is shown. *Bottom:* During the day, clock (CLK) mRNA and protein levels decline, whereas VRI mRNA and protein accumulate. In the light, CLK switches on its natural repressor VRI. This simplified summary does not include every autoregulatory link known and emphasizes transcriptional feedbacks rather than the post-transcriptional steps.

Mouse clock (*right*). *Top:* Negative feedback loops at the transcriptional level coupled with time delays make a rhythm. R = orphan nuclear receptor REV-ERB-alpha. *Middle:* During the day, REV-ERB-alpha restricts the BMAL protein product's function, thus keeping PER/CRY activity low. *Bottom:* In the morning, high BMAL protein levels induce a wave of *Per*, *Cry*, and *REV-ERB-alpha* transcription. Accumulation of *Per* and *Cry* proteins will be delayed until after dark. Note that other autoregulatory steps are already known in the literature, and that this simplified adaptation emphasizes the transcriptional mechanisms in the mamalian circadian clock.

Adapted from Young and Kay (fly clock: *Nat Rev Genet*, 2001; 2:702–15) and Young (mammalian clock: *Neuron*, 2002; 36:1001–5).

ity of enzymes, very different chemistries—different classifications among the chemical groups—are involved, and this variation provides even more safeguarding. A single biochemical deficit cannot wipe out the system. Finally, the different kinds of genes supporting arousal provide for many different types of regulation. Indeed, these several forms of regulation allow the fluctuations, highly variable in their timing and amplitude, that give the arousal systems their high-information content. In keeping with one of the main themes of this book (Chap. 1), neural systems for arousal not only respect environmental situations and inputs with high-information content (Chap. 1); they also exhibit high-information fluctuations by virtue of their own lability.

2. Heterogeneity providing flexibility of response. If the very multiplicity of the gene products bearing on arousal raises the essential information content exhibited by neural systems for arousal, the differences among these genes' functions heightens the possibilities even further. Not all the genes covered in this chapter pull in the same direction. In particular, there is a tendency for gene duplication products and differential splicing products to have much different effects on electrophysiologic or behavioral responses.[518] Just four examples will suffice. DA receptors 1 and 5 differ markedly in their signaling and biophysical functions compared to DA receptors 2, 3, and 4. Transcriptional functions exerted by ER-α can oppose those exerted by ER-β. Behavioral effects of TR-α contrast with those of TR-β.[519] The nine–amino acid peptide oxytocin has behavioral functions that can oppose those of its gene duplication product, vasopressin.

3. Directionality lending to acceleration. The genes reviewed in this chapter provide the mechanisms for sudden increases in transcriptional systems that contribute to increased arousal and for decreases in those that cause decreased arousal. This arrangement tells us how we could achieve a rapid change of CNS state during environmental circumstances when a sudden increase in the arousal of brain and behavior is required. Positive feed-forward mechanisms, flagged at several points in this chapter, are potential mechanisms for the acceleration.

This insight immediately raises the question: Once you have achieved a high state of alertness, perhaps for an emergency, how do you turn it off? I address this question from a neural engineering point of view in Chapter 7.

4. Stable patterns forming temperaments. Long-lasting behavioral tendencies sometimes have genetic predispositions. Huge inheritable strain differences in mice, rats, dogs, horses, subhuman primates, and other animals attest to the impact of genetic contributions on arousal-based behaviors. For human psychology, consider mood differences between genders that are attributable to the Y chromosome. What do we think about genetic contributions to human behavior?

Genetically trained psychiatrists have begun to discern traits of human behavior that show clear genetic influences.[333,334,520] The tendency for men to have depression less frequently than women may have something to do with the Y chromosome; but beyond the Y, other chromosomal regions contributing to depression in women have been identified.[521] Animal models of depression offer the possibility of figuring out which drugs will be most effective for patients with particular gene polymorphisms.[522] Genetic differences also contribute to individual differences in behaviors related to attention,[523] intelligence tests,[524] and autonomic nervous system function.[525] In sum, our growing knowledge of behavior genetics will increase our understanding of normal behavior[526] and our skill at using pharmacology to improve abnormal behavior.[527]

There are still more implications of there being so many genes controlling arousal mechanisms. The very multiplicity yields the possibility of large numbers of meaningful *patterns* of gene expression. In a neuroendocrine context, we have shown that it is impossible to understand gene/behavior relations on a one-by-one basis. Charging beyond Beadle and Tatum's classical concept from their work with the fungus Neurospora—"one gene/one enzyme"—we have reached the conclusion that different *patterns* of gene expression yield different *patterns* of sociosexual behaviors.[528]

Reasoning further, the multiplicity of contributing genes and the stable patterns of expression also provide a lot of room for seeing how individual differences in temperament come about. We have the genetic capacities and variety to produce an incredibly large number of personality differences. Moreover, there is a more subtle type of chemical modification to consider, as far as real-life applications of medical genetics are concerned. When we think about differences between individual humans, we must move beyond the gross genetic experiments in animals in which genes are simply deleted and consider the real-life situation with humans in which a variety of single-nucleotide base mutations can alter, very slightly, the coding region in a gene or the quantitative efficiency of its promoter. These more subtle differences provide a seemingly infinite number of ways in which human feelings and dispositions can be fine-tuned. I propose that these small genetic variations help to explain large numbers of individual differences within the universe of human temperaments.

■ Summary

At present count, at least 124 gene products participate in the regulation of arousal in the mammalian brain. The participation of some of these genes was predictable because they are involved in classical neuroanatomic systems controlling the arousal state. The involvement of others, such as orexin/hypocretin, has only recently been recognized and is incredibly important. The number of the genes involved not only attests to the importance of the state functions covered in this book but also provides for state changes that are reliable, flexible, and sometimes rapid.

6 Heightened States of Arousal: Sex Compared to Fear

Two emotional states, sex and fear, highlight two extreme types of behavioral response, approach and avoidance, respectively. Both states have been analyzed mechanistically and both encompass generalized arousal together with specific arousal conditions. Unpredictabilities—high-information conditions—elevate the thrill of sex and intensify the grip of fear.

C an arousal concepts illuminate the mechanisms for specific biologically regulated behaviors? This chapter applies these concepts to a behavioral system that we have worked out. For mating behaviors, we know the most about how environmental and hormonal stimuli alter genomic and biophysical mechanisms in the CNS, thus triggering a behavioral response. With that knowledge as background, we explore potential "trading relations" between a specific form of arousal (sexual) and generalized arousal. The equation in Chapter 1 promises that new data will lead to a rational, mathematical structure of arousal. You can find the beginning of this effort in the section "Generalized Arousal Affects Specific Arousals and Vice Versa." Then, we compare sexual arousal to another heightened state, fear. In this comparison I revive the classic ethological division of behavioral responses—approach versus avoidance. Any given elevation of arousal level could fuel either type of response, depending on the animal's (or human's) assessment of risk and initial state of mind.

Information theory concepts, once again, add both precision and insight to the mechanisms by which these behaviors operate, as we explore near the end of the chapter. Uncertainty about the near future influences both fear and sex response mechanisms. Finally, I claim that arousal mechanisms—hormonal neural and genetic—impact sex and fear in humans as well.

I chose to compare these particular neurobiologic systems, sex and fear, because of our depth of knowledge in mechanistic terms about how they work. They represent only a small part of an exploding field, neuroendocrinology.

The breadth of this field has been represented in a large reference source[529] and systematized in a didactic fashion.[518]

■ Sex Behavior's CNS Mechanisms Require Arousal

For many reasons, sex behavior mechanisms were easier to figure out than many other natural, biologically regulated behaviors. There were, for example, strategic advantages to analyzing responses controlled by steroid sex hormones. Working with steroid hormones gave us all the advantages of a well-developed steroid chemistry, isotopically labeled compounds in days when such were barely available, a burgeoning pharmarmacology (anti-estrogens, anti-androgens, and so on) and all the tools of molecular endocrinology. Mating responses have a relatively simple behavioral topography. They are also biologically crucial behaviors that can be evoked in their natural form in the laboratory. Finally, strong hormone/behavior relations are demonstrated in many species, including higher primates[530] and humans (C. S. Carter in Schmitt et al.[531]). What differs among species is the share of the "causal pie" due to hormonal stimulation as compared to the influences of culture and context.

As a result of all these strategic advantages, we have known sex behavior CNS mechanisms for a long time. In the 1960s we could see that sex hormones circulating in the bloodstream are received, strongly bound, and retained for a long time by neurons in a limbic-hypothalamic system of neurons.[532] In addition, however, we could detect some hormone retention in the CNS far from this limbic-hypothalamic system. Then the neurons in one specific group of hypothalamic neurons were discovered to be at the top of a neural circuit governing a simple female-typical sex behavior, an estrogen-dependent behavior called lordosis (Fig. 6.1). In four-footed female animals, this primary mating response is so simple that its mechanism could be revealed[19] as the first nerve cell circuit understood for any vertebrate behavior.

The hormone receptors discovered in the brain—so-called ligand-activated transcription factors—turned out to be nuclear proteins capable of facilitating gene expression. Sex hormone effects on reproductive behaviors are blocked by RNA and protein synthesis inhibitors. Thus, we could use molecular endocrine techniques over a period of years (summarized in Pfaff[20]) to discover genes that have two properties: they are turned on by estrogens, and their products foster female reproductive behaviors (Fig. 6.2). The list of genes in Figure 6.2 leads to at least two questions.

First question: can we conceive of the biologic importance of these facilitations of genomic transcription in a systematic fashion? Yes, Figure 6.3 shows

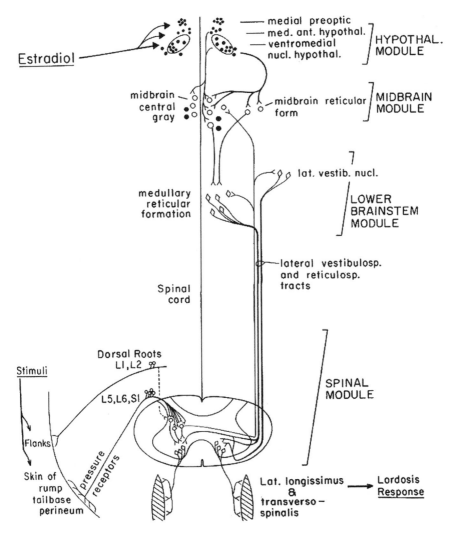

Figure 6.1. The neural circuit that produces lordosis behavior, the primary estrogen-dependent sex behavior in female four-footed mammals (from Pfaff, *Estrogens and Brain Function*, 1980). The circuit is bilaterally symmetric, but is drawn here on one side only, for visual clarity. The behavior is triggered by cutaneous stimuli, facilitated by estrogenic action in ventromedial hypothalamic neurons, and is manifest in massive contractions of the deep back muscles (*bottom*). Its neuroanatomic and neurophysiologic features indicated a modular construction (*right*), which turned out to match embryologic divisions of the neuraxis.

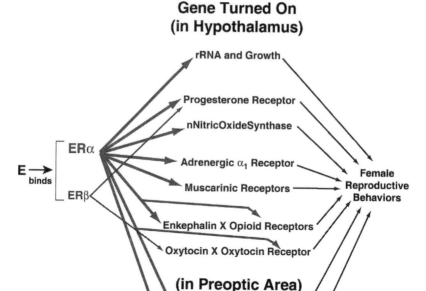

Figure 6.2. This list of genes has two properties: (1) estrogen (E) (estradiol) treatment raises their transcript levels following E binding to estrogen receptor (ER)-α or -β; and (2) their products facilitate lordosis behavior. A microarray study revealed an exception: Prostaglandin-D synthase mRNA is *reduced* in the preoptic area, where it *inhibits* lordosis behavior. Their concerted actions currently are conceived as functional genomic modules downstream from hormone action (see Fig. 6.3). From Pfaff, *Drive*, 1999.

---

modules of estrogen-facilitated genes and functions that serve reproduction in a biologically sensible fashion. They include the hormone-induced growth of neurons, the *a*mplification of the estrogen effect by progesterone, the *p*reparative behaviors induced, the *p*ermissive effects of hormone actions on hypothalamic neurons; and the *s*ynchrony of sex behavior with ovulation in lower animals (GAPPS).[533] Coordinated reproductive behaviors synchronized with endocrine preparations for reproduction emerge through the concerted actions of these genetic modules (Fig. 6.3).

Second, is the list of genes in Figure 6.2 curtailed by lack of time, imagination, or reagents? Yes, again. We have solved that problem with microarrays of oligomers that let us estimate the hormone sensitivity of more than 11,000 genes in a single experiment.[483,512] One of the early discoveries of a novel

MODULAR SYSTEMS DOWNSTREAM FROM HORMONE-FACILITATED TRANSCRIPTION RESPONSIBLE FOR A MAMMALIAN SOCIAL BEHAVIOR: "GAPPS".

Growth (rRNA, cell body, synapses).
Amplify (pgst/PR $\longrightarrow \longrightarrow$ downstream genes).
Prepare (indirect behavioral means; analgesia
      (ENK gene) and anxiolysis (OTgene).
Permit (NE alpha-1b; muscarinic receptors).
Synchronize (GnRH gene, GnRH Rcptr gene
      synchronizes with ovulation).

Figure 6.3. Genes from the list in Figure 6.2 are arranged in the form of functional modules: the GAPPS model. (From Mong and Pfaff, _Mol Psychiatry_, 2004; 9:550–6.) Abbreviations: ENK, enkephalin, an opioid peptide; GnRH, gonadotropin-releasing hormone, controls reproductive physiology; OT, oxytocin, a hypothalamic neuropeptide; PR, progesterone receptor, binds progesterone and is itself a transcription factor.

---

estrogen-dependent gene from a microarray turned out to be important for linking generalized arousal with mechanisms of sexual arousal, prostaglandin-D synthase.

Sex hormone actions in CNS play out not only in terms of behavior but also in terms of the electrical activity of neurons. New experiments using electrophysiologic approaches with the patch clamp technique permit us to analyze sex arousal mechanisms in cells identified individually in five ways: by location, size, shape, mRNAs expressed (we suck out the cytoplasm to measure these after recording from the cell), and behavioral importance.

The functional genomics of sexual behaviors as well as aggressive behaviors have been pursued most easily by using animals with gene knockouts. Over a period of years we have been privileged to work with Jan-Ake Gustafsson at the Karolinska Institutet and Ken Korach at the National Institutes of Health to discern the phenotypes of estrogen receptor-α (ER-α) and estrogen receptor-β (ER-β) gene knockout mice. ER-α is responsible for an entire chain of events intimately connected with nitty-gritty reproductive physiology. Estrogens are secreted by developing follicles in the ovaries of a woman or a female animal who soon will ovulate. These hormones circulate in the blood, enter the brain, and are concentrated and retained by cells in the hypothalamus and preoptic area.[19,532] Their molecular endocrine signaling through ER-α is required for normal courtship behaviors in our female mice, then for lordosis behavior, and hence for normal maternal behaviors.[534–536]

Thus, even as estrogens prepare several organs in the body of the female mammal for successful reproduction, they set up the CNS for the corresponding behaviors. Estrogens prime the pituitary gland for releasing the ovulatory surge of protein hormones and they control the ovaries and uterus themselves. At the same time, estrogens working through ER-α trigger events in the hypothalamic and preoptic neurons that control behaviors on a timeline corresponding to the other reproductive-physiologic events in the body. This chain of behaviors stretches all the way from courtship behaviors—locomoting toward the male, the eventual father—through sex behavior, and then through the proper caring for young babies.

In dramatic contrast, ER-β is not required for lordosis and, in fact, inhibits lordosis.[537] More generally, as Gustafsson has worked out, ER-β often opposes ER-α actions.[538] In the brain, ER-β is required for normal social recognition and memory (Choleris et al.[539]; see Fig. 1.7). In mice, these in turn are necessary for a range of affiliative behaviors and for the inhibition of aggression.[540]

Looking to the future, functional genomics will not depend entirely on gene knockouts. Now, improved antisense DNA approaches[541] and RNA interference (RNAi) applied to neurons, viral vectors, and sophisticated applications of molecular pharmacology allow us to block behavioral mechanisms in the CNS in a manner that is specific temporally and neuroanatomically.

In summary, this huge body of knowledge about genetic and brain mechanisms for simple sex behaviors provides a "launching platform" for analyzing sexual motivation and arousal, and then generalized arousal. Hormone-dependent sexual arousal gets us into the equation (Chap. 1), linking specific and generalized arousal mechanisms. The first step in this expansion is to realize that hormone effects on sex behaviors prove, in formal terms, the existence of an underlying sexual motivation that is elevated by hormonal treatment.

Sexual Motivation. Suppose you have an animal, lacking any sex hormones, in a well-controlled experimental environment. You supply a stimulus animal of the opposite sex for mating and nothing happens. The test animal does not mate. Then, suppose you inject the test animal with an appropriate sex steroid hormone and retest—same controlled environment, the same time of day, the same age of test animal, and the same stimulus animal. The test animal mates. In the logical equations that describe behavior, the stimulus and the response have been held constant. Therefore, the sex steroid hormone must have altered another variable term in the equation. That variable is called sexual motivation. The authority Charles Cofer[12] calls sexual motivation "the most powerful factor in energizing and directing behavior" (p. 175). It depends on the incentive value of the potential sex partner[542] and the sex drive of the test animal (or human).

Remember the analogy in Chapter 1. If behavior were viewed as a vector and the incentive object determined the angle of the vector, motivation dependent on arousal determines the length (amplitude) of the vector. Motivational concepts are absolutely necessary to explain a wide variety of biologically regulated responses in animals and humans.[12] Logically, they have the same status in behavior-CNS equations as gravity has in Newton's second law.[18] Explaining mechanisms for a motivational state presents a real gift to experimenters: We would be explaining entire classes of behavioral changes—changes in the state of the brain—rather than working hard just to explain an individual behavior. For all these reasons, neurobiologists, behavior analysts, and ethologists have historically embraced motivational concepts.

Even in simple mammals such as mice, motivation to mate plays out in a series of courtship responses. Given an adequate environment, for example, a "seminatural environment" as constructed in our lab,[543] the female will initiate unusually fast and directed locomotion to orchestrate her contact with the male. In fact, an entire series of communicative and locomotor behaviors culminate in approach responses. Social investigation and affiliative behaviors are increased, and aggression decreased. An interesting side point of mouse locomotor activity is that it is strongly rhythmic according to the daily light cycle, thus allowing the experimenter to play "circadian games." A more important point: In the female mammal these courtship and locomotor behaviors depend very much on circulating estrogen levels[29,30,544–546] working through ER-α.[546] More generally, a wide variety of behavioral assays show that a female animal's sexual motivation is estrogen-dependent (reviewed, Pfaff[547]). For locomotion, estradiol enters the brain, is bound by neurons in the preoptic area,[532,548] and through those preoptic area neurons facilitates locomotion by the female.[549]

What happens in the preoptic area to allow estrogens to increase running? Estrogens excite electrical activity there,[550] specifically in neurons connected with locomotor circuits.[551,552] The neurons in the region identified by Sakuma and his laboratory at Nippon University in Tokyo[553] are associated with a preoptic locomotor region,[554,555] which likely works through a midbrain locomotor region.[556] In fact, Kato and Sakuma[557] were able to correlate electrical activity in certain of these preoptic neurons with the initiation of courtship behaviors by female animals.

What genes are necessary for estrogens to turn on locomotor behavior? Clearly in female mice, the gene that encodes estrogen receptor-α is necessary (Fig. 6.4), and the gene that encodes estrogen receptor-β is not.[546]

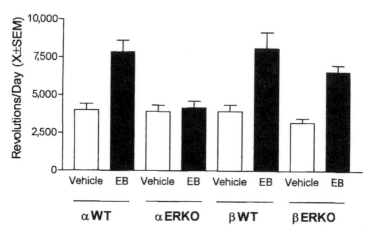

Figure 6.4. Estrogenic effects on locomotion in female mice depend on the gene for estrogen receptor-α, but not the gene for estrogen receptor-β. (From Ogawa et al., *Endocrinology*, 2003; 144(1): 230–9). In mice with estrogen receptor-α knocked out (α-ERKO), there was no effect of estradiol benzoate (EB, black bars) on running wheel activity compared to the same animals given vehicle control (open bars). However, their wildtype littermates with the same genetic background (α-WT) did show the expected effect of estradiol on raising voluntary locomotor activity. In contrast, the estradiol effect was robust in animals in which the gene for estrogen receptor-β had been knocked out (β-ERKO), as well as in their wildtype littermate controls (β-WT).

While a tremendous amount of endocrine and behavioral experimentation has been done with typical laboratory animals such as rats and hamsters and genetic work has focused on mice, the same themes show up in work with primates such as rhesus monkeys. Kim Wallen and his colleagues at Emory University have demonstrated in both female and male monkeys that gonadal steroid hormones can elevate sexual motivation manifest in various courtship and approach responses.[530] In the female, these require ER gene products, because they can be blocked by an anti-estrogen[558].

Sexual Arousal. In the literature on mechanisms of motivation, it is clear that the occurrence and forcefulness of any motivated behavior by animals or humans depends on arousal. Therefore, all of the knowledge summarized in Chapter 1 informs our discussion here.

Historically, neurobiologists were influenced to consider heightened arousal as a requirement for any significant motivational state by the studies and arguments of Donald Hebb,[13] who incorporated neurophysiologic and neuroanatomic features of the ascending reticular activating system into his theory of motivation. For the strong muscular contractions and use of metabolic energy exerted during vigorous forms of motivated behavior, animals or humans could not get along without high levels of arousal. A perfect example is the rapid locomotion of female rodents during courtship behaviors. Such locomotion depends upon estrogens. In human psychology, as well, especially in motivational psychology, you cannot escape concepts of arousal. When constructing theories of personality, authors always invoke the dimension of arousal together with the dimension of valence. Therefore, it is a smooth logical and scientific road we travel as we go from explaining concrete sexual behaviors, to sexual motivation, to arousal.

Importantly, sexual arousal with its specific molecular endocrine triggers and its simple behavioral endpoints provides a tractable scientific approach to the dynamics of arousal mechanisms as they impact all behaviors. Sexual arousal is simply one example of a specific arousal force that joins together with generalized arousal (see Chapter 1) to enable reproductive behavior. Now, we can begin to explore the internal structure of the arousal equation in Chapter 1 by looking at the trading relations between sex arousal mechanisms and generalized arousal.

■ Generalized Arousal Affects Specific Arousals and Vice Versa

Relations between generalized CNS arousal states and specific forms of arousal are a two-way street. In this section, we treat both, in order.

Effects of Elevated Generalized Arousal on Sexual Arousal

In both female and male experimental animals, neurochemical manipulations that mimic ascending arousal systems increase sexual behavior. Three of the neurotransmitters give especially clear stories: noradrenaline (NA), histamine (HA), and acetylcholine (ACh), as studied in female experimental animals. After these three, I will treat a novel gene involved—the one coding for prostaglandin-D synthase.

NA. NA administration to female rats increases the primary female sex behavior, lordosis, and also mediates the ovulatory surge of luteinizing hormone (LH) from the pituitary(reviewed, Etgen[559]). Thus it helps to synchronize the female's sexual arousal with her endocrine preparations for reproduction. Conversely, depleting hypothalamic NA abolishes lordosis behavior and disrupts the LH surge. The NA effects in the hypothalamus work through adrenergic α-1b receptors.[560,561] They originate with NA cell bodies in the lower brainstem, following the "low road" (Chap. 2, e.g., from locus coeruleus; Helena,

Franci, and Anselmo-Franci[562]). They work by increasing the electrical activity of the ventromedial hypothalamic cells that control lordosis (Fig. 6.5) and of the preoptic cells that control the locomotion of courtship behaviors. Reducing ventromedial hypothalamic neuronal electrical activity—for example, by using a high dose of a selective μ-opioid receptor agonist—correspondingly reduces lordosis.[563] In addition to the straight electrophysiologic effects of estrogens promoting lordosis behavior, Etgen and her colleagues[560] have implicated growth factor signaling in these hormonal actions. Exactly how these growth factors interact with neurotransmitter control over sexual arousal remains an open question.

HA. This neurotransmitter, produced in the posterior hypothalamus, is intimately involved in arousing the entire forebrain (Chap. 2). It also increases sex behavior, or lordosis, in experimental animals. One mechanism by which HA acts is that it increases electrical activity in the very hypothalamic cells that control the lordosis behavior circuit (Fig. 6.5).

ACh. Cholinergic actions in the hypothalamus promote lordosis behavior (reviewed, Dohanich, McMullan, and Brazier[564]). As you would expect, one way they can do this is by increasing the firing of action potentials by ventromedial hypothalamic cells (Fig. 6.5). In addition, acetylcholine biochemistry ties into the mechanisms by which estrogens facilitate female sex behavior. Estradiol treatment not only increases the activity of the rate-limiting enzyme for ACh biosynthesis, choline acetyltransferase, but also increases the receptors through which ACh works.[559]

Prostaglandin-D synthase (PGDS). Our logic works in the other direction as well. Inhibiting a neuromodulator that decreases arousal will heighten sex behavior. Estrogens decrease transcript levels for prostaglandin-D synthase in the preoptic area.[486] Experimentally reducing PGDS specifically in the preoptic area by the use of a novel antisense DNA microinjection approach increased lordosis behavior.[483]

Thus, in females, neurotransmitters acting to increase generalized arousal also increase sex behavior. What about in males? For male sex behavior the focus shifts from the ventromedial hypothalamus to a cell group in the basal forebrain just in front of the hypothalamus—the medial preoptic area. The arousal-related transmitter now taking center stage is dopamine (DA).[116] In male rats, DA increases male sex behavior through at least three functional roles. It increases sexual arousal and the courtship behavior that goes with it. It also potentiates the motor acts of mounting behavior. Finally, it facilitates gen-

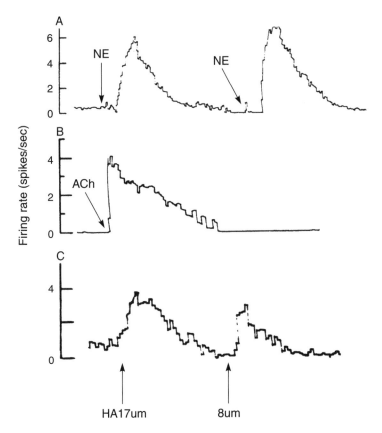

Figure 6.5. Neurotransmitters signaling generalized arousal can elevate electrical activity in ventromedial (VM) hypothalamic neurons responsible for sexual arousal and sexual behavior. Recordings here are from VM hypothalamic neurons in female rats. Illustrated are examples of responses to (A) norepinephrine, NE; (B) acetylcholine, ACh; and (C) histamine, HA. From (a) Kow and Pfaff, *Brain Res*, 1987; 13:220–8; (b) Kow and Pfaff, *Brain Res*, 1985; 347:1–10; (c) Jorgenson, Kow, and Pfaff, *Brain Res*, 1989; 502:171–9.

ital responses to stimulation. Testosterone promotes DA release in the basal forebrain and this release, in turn, is timed to coincide with actual mating behavior by the male rat. You will recognize that the impact of generalized arousal on sexual arousal is even stronger when you appreciate that erections and ejaculations require complex activation of both the parasympathetic nervous system and the sympathetic nervous system, also intimately involved in CNS arousal (Chap. 4). In fact, a hypothalamic cell group crucial for autonomic nervous system control, the paraventricular nucleus, sees more dopamine released during male sexual activity. Other arousal-related neurotransmitters tied to male sexual performance include NA[93] and glutamate.[565] Most exciting is the orexin gene product (Chap. 5). Microinjecting orexin into the preoptic area of male rats potentiates their sex behavior.[566] In sum, several neurochemicals that heighten generalized arousal also foster male sexual arousal.

Do not think that these demonstrations of close relations between generalized arousal and sexual arousal are limited to experimental animals. In heterosexual men, sexual interest is strongly suppressed during depressed mood states.[567] Further, in women and men, drugs used to treat decreased libido include those that operate on generalized mood and arousal states (reviewed, Lopez and Koller[568]).

Effects of Elevated Sexual Arousal on Generalized Arousal

Generalized arousal mechanisms depend on sex by virtue of sex hormone binding in generalized arousal neurons. Estrogenic hormones are accumulated in NA cell groups in the lower brainstem that give rise to major ascending arousal influences.[95,209,548,569–573] The novel estrogen receptor gene product, ER-β, is found in the midbrain raphe nuclei[148,574] giving rise to serotonin pathways, as well as in midbrain cells giving rise to DA pathways.[575] We can be sure that in the nerve cells where these sex hormone receptors are found, they are affecting gene transcription,[528] membrane biochemistry and electrophysiology,[576,577] or both.[578]

A completely different mechanism for sex to impact generalized arousal depends on sensory physiology. Sexually significant stimuli from the genitalia, coming into the body over the pelvic and pudendal nerves, signal strongly to the lower brainstem reticular formation's nucleus gigantocellularis,[99,196,579,580] which in turn affects locus coeruleus,[196] which then wakes up the entire forebrain. These are huge neurobiologic phenomena represented in the several species studied. Potent sexual influences thus impact the very master cells (Chap. 2) that form the core of generalized arousal mechanisms. Through all of these molecular, electrophysiologic, and neuroanatomic connections, generalized arousal pathways are jazzed up through sexual behavior mechanisms.

Going deeper into the molecular biology of sex hormone action, we can see how sex hormone effects in brain contribute to generalized arousal. The molecular routes of influence are almost too numerous to count. Prominent among these are the ways in which sex steroids ramp up NA synthesis and effectiveness. Estradiol stimulates gene expression for the enzymes that synthesize NA in that important cell group in the brainstem, the locus coeruleus.[413] Then, in the hypothalamus, estrogens stimulate gene expression for a specific

NA receptor subtype, the α-1b receptor,[581] and foster interactions between these receptors and signal transduction pathways in hypothalamic neurons.[582] In the basal forebrain, NA α-1 receptors cooperate with NA β-receptors to increase arousal[583] by inhibiting sleep-active neurons and exciting wake-active neurons.[584] Sexual interactions by female rats with males evoke NA release in specific parts of the hypothalamus.[411] Thus it appears that a sex hormone increases the signal-to-noise ratio in arousal-related pathways by acting at an entire train of mechanisms, from synthesis through release through receptors through postsynaptic action.

Estrogens work through several other arousal-chemical systems as well. DA is an interesting case. Ingrid Reisert at the University of Ulm, Germany, has found that sex steroids can promote neurite outgrowth in midbrain DA neurons.[585,586] Many of the estrogenic effects on DA neurons appear to be on the output side. They affect the amount of time DA can reside in the synapse for greater effectiveness[587] and the effects of DA on signal transduction pathways in the postsynaptic neuron.[588] Estrogen receptors are also found in many histaminergic neurons.[589] Indeed estrogen (E) administration can amplify neuronal responses to histamine.[590] Another arousal-related neurotransmitter, serotonin, is susceptible to estrogenic effects. E heightens gene expression for the rate-limiting synthetic enzyme tryptophan hydroxylase[591-593] and also affects serotonin receptors. Serotonin is tricky because of the tremendous variety of its receptor genes (fourteen), and because of its autoreceptors, presynaptic receptors that turn down serotoninergic transmission. Sex hormones desensitize the 5-HT1A autoreceptor,[594,595] thus facilitating serotonin's synaptic functioning and, probably, heightening mood. Sex hormones heighten cholinergic function not only in the hypothalamus[596] but also, importantly, in the basal forebrain.[597,598]

The latter is important for arousal because of the widespread distribution of axons from these cholinergic neurons throughout the cerebral cortex. Finally, regarding conventional neurotransmitters, it is only fitting the glutamate receptors—likely transmitters of some of the master cells (Chap. 2)—play a role in estrogenic effects on autonomic arousal.[599] Altogether we have covered six neurotransmitters that work to convey specific sex hormone effects on generalized arousal.

Neurochemicals that are not long-recognized transmitters also carry sex hormone influences on arousal. Orexin receptors (Chap. 5) are, in some locations, susceptible to sex hormones.[600] Neuronal nitric oxide synthase—manufacturing the gaseous compound nitric oxide and elevated by estrogens in the hypothalamus[601,602]—is important for diffuse cerebral cortical electrical activation.[603] A peptide central to stress and fear recognition, CRH (corticotropin-releasing hormone), has its transcript levels significantly elevated by estradiol in hypothalamic neurons.[604]

A powerful, newly discovered molecular mechanism by which estrogens elevate generalized arousal popped out of a microarray (a cRNA chip). In the basal forebrain—in particular, in a cell group in the lateral preoptic area that controls sleep—estradiol administration significantly *reduced* gene expression for prostaglandin-D synthase (PGDS),[483] the enzyme that synthesizes prostaglandin D. Now, if you put prostaglandin D in this same part of the basal forebrain, you put the animal to sleep. Using a new technique called antisense DNA with so-called locked nucleic acids, Jessica Mong[486] reduced the expression of PGDS mRNA experimentally and thus increased arousal level. In addition, reducing expression for PGDS also permitted sex behavior, even in the absence of sex hormones.[486] Through this genomic route, sex hormones worked on a mechanism for generalized arousal that then heightened sexual arousal.

The PGDS finding also explains another effect of estrogens in the brain. E administration reduces sleep duration, especially that deep state called slow-wave sleep.[605-607] These are classic findings in experimental animals. On the other hand, perimenopausal women given estrogen report decreased latency to sleep onset and increased total sleep time (reviewed, Manber and Armitage[608]). Whether this beneficial effect reflects hormone replacement therapy (HRT) patients simply feeling better remains to be determined.

As a side point, when we talk about an individual being "hot" with sexual arousal, we usually do not mean it literally. However, Kiyatkin and Mitchum[609,610] have discovered that sexual activity does indeed raise an animal's brain temperature, and that there are differences in this phenomenon between males and females.

In summary, through several molecular and neurochemical routes, a sex hormone like estradiol is able to increase generalized arousal. Sex hormone influences in the brains of experimental animals encourage robust sexual interactions. These interactions actually raise brain temperature. Sexually "hot" literally makes for brain heat.

We can now address new theoretical questions such as, How do we quantify the "trading relations" between generalized arousal and sexual arousal? Could there be positive feedback between generalized arousal and sexual arousal? Chapter 7 explores these questions in detail. In some individuals, the impact of sexual stimuli on overall arousal might be much stronger than the loading of generalized arousal onto sexual excitement. In other individuals, is

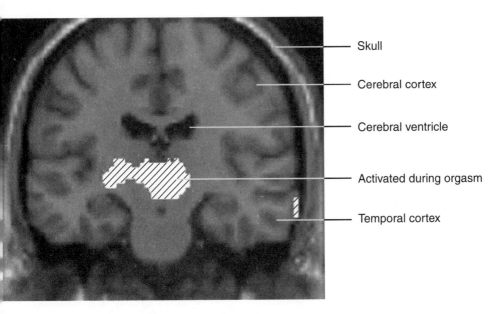

Figure 6.6. Gert Holstege and his colleagues at the Medical School, University of Groningen in the Netherlands, have done brain-scanning experiments during human male ejaculation. Among other areas activated, an especially interesting one, represented above, is in the ventral tegmental area, which has been associated with reward. Similar experiments concentrating on visualization of beloved partners were done by Lucy Brown (Albert Einstein Medical School) and Helen Fisher (Rutgers University; see Society for Neuroscience Abstracts, 2003, p. 725.27; and *J Neurophysiol*, 2005). (From Holstege et al., *J Neurosci*, 2003; 23:9185–93.)

the reverse true? As introduced in Chapter 1, the mathematics of arousal and the structure of its mechanisms are ready for investigation.[27]

From all of the findings outlined in this chapter, it is easy to see how sex hormones affect mood states in both men and women. Estrogens heighten mood in certain groups of women.[611,612] Androgens have the same effect, especially in older men. These data fit with effects of estrogens in experiments intended to mimic depression in rats[613-615] and mice.[616,617] They also agree with effects of estrogen levels on cortical activation in women.[618] Love associated with adequate sex hormonal support is a primordial, powerful emotional state, a kind of arousal associated with reward (Fig. 6.6); and the physical anthropologist Helen Fisher has shown that it activates large parts of the subcortical forebrain.[619] In fact, Deborah Grady, a professor of medicine at the University of California (San Francisco), has been quoted as saying that estrogen administration makes postmenopausal women feel so good that it is almost addictive.

Speaking of addiction, Jill Becker and her laboratory at the University of Michigan have demonstrated with several experimental strategies that estrogens enhance behavioral sensitization to cocaine.[620,621] These effects certainly work through altered DA release in the basal forebrain, and they may also work through altered gene expression in the brain.[622] From Becker's work, it appears that estrogenic enhancement of drug effects comprises still another route for sex hormones to facilitate generalized arousal.

Do Effects of the Initial Condition of the Animal Permit Hormones to Optimize Sexual Motivation?

Some of the data covered in this chapter raise the possibility that arousal levels are fine-tuned in a way that facilitates sex. This theory comes in three parts. First, there absolutely must be an optimal level of arousal for reproduction (Fig. 6.7). Consider the extreme low on the arousal scale. If an animal or human being was comatose, stuporous, or sluggish, he or she would never "get it together" to manage courtship and mating. What about the extreme high side? If the animal or human was absolutely frantic, terrified, or delirious, he or she

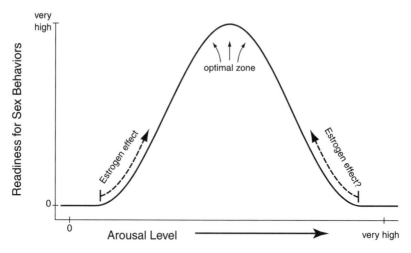

Figure 6.7. Is it possible that for certain hormone/brain/behavior relationships, the nature of the hormone effect depends on the initial condition of the animal? Sketched here diagrammatically is one fact about arousal that we know for sure along with one unexpected theoretical possibility. Certainly, there is an optimal level of generalized arousal for sexual arousal to be expressed in sexual behaviors (see text). In addition, might the nature of the estrogen effect depend on the initial level of arousal, from which the animal or human starts? Coupled with this question is the possibility that hormonal influences acting through altered arousal levels affect specific behaviors in a manner that depends on whether the environment is frightening (see Fig. 6.8).

would be too upset and fearful to make love. Between the extremes of very low arousal and very high arousal, the curve must reach a peak. This is the optimal level of arousal for mating.

Second, suppose that the effect of a hormone depends on the initial condition of the animal or human being (i.e., the condition just before the hormone is given). I pose this idea because the following section shows that the effect of estrogens on arousal can be manifested as increased activity or increased fear (i.e., decreased activity) in mice, depending on the environment. Already we know that if the initial arousal condition is very low, estrogens will raise it. Suppose that estrogens would lower extremely high arousal conditions (Fig. 6.7). Then we would get a very interesting result, as explained in the third point.

Third, because there is an optimal level of arousal for reproduction, and assuming estrogens push arousal levels from the extremes toward the middle, estrogens optimize arousal level for reproduction. This theory needs systematic testing.

If we extrapolate to the strongest of human feelings, Figure 6.7 may embody a neurobiologic theory of euphoria. When we are soporific or down, sex hormones excite us. When we are overstimulated and too anxious, according to this idea, sex hormones relax us. As we approach the optimal level of arousal, a warm feeling of well-being overwhelms us. True? Only new observations will tell.

■ Contrasting Sex and Fear

I use sexual arousal and behavior as an example because, among motivated behaviors, mechanisms for sexual arousal are best worked out (see the previous section "Sex Behavior's CNS Mechanisms Require Arousal"). Another powerful motivator is pain. The neural pathways ascending to the brain and signaling pain, and descending from the brain and limiting pain, run parallel to the neural pathways involved in sex.[18] Sex differences in sexual behaviors are well understood. Sex differences in pain and its control[18,623] are well established, but most of their mechanisms are still obscure.

The anticipation of pain strikes fear in our hearts. If there is any emotion that grabs our minds as completely as sexual desire, it is fear. Until very recently, this subject was ignored by neuroscientists. It did not have the same molecular biologic cachet as hormone-driven sex behaviors, and it often had to be studied by artificial fear-conditioning procedures. Now the neuroanatomy of conditioned fear has become clearer (reviewed in a number of studies[367,624–626]). Basically, stimuli connected with fear learning must reach the amygdala, an almond-shaped group of neuronal centers in an ancient part of the basal forebrain. In a variety of studies on experimental animals from rodents to primates, as well as in work with human patients, the amygdala has been linked to emotions, including fear. Data from Joseph LeDoux's lab at New York University suggests that conditioned stimuli for fear must enter a cell group called the lateral amygdala. The lateral amygdala communicates with an output cell group called the central amygdala, which sends signals to other parts of the CNS responsible for organizing fear responses. LeDoux[627] sees the system as designed for speed. Stimuli signaling fear travel parallel pathways, some with crude resolving capacity but very fast, others with more detailed information but slower. The operations of these pathways may not be exactly the same in different individuals. In fact, inherited differences between or within species are likely to reflect a genetic contribution to fear behaviors[627] (pp. 134–137).

One obvious possibility for genetic influence has to do with the neuropeptide primarily in charge of organizing the body's endocrine response to a fearful stimulus, which was discovered by Wylie Vale at the Salk Institute. The expression of corticotropin-releasing hormone (CRH) by neurons in the paraventricular nucleus of the hypothalamus is crucial for the control of stress hormones by the pituitary gland; but CRH is also expressed by many other regions of the brain, notably the amygdala (reviewed, Valentino and van Bockstaele[492]). Studies with mice whose gene for producing CRH has been deleted[628] suggest that not only CRH but also newly discovered CRH-related peptides work through two receptor systems, CRH-1 and CRH-2. These facilitate responses to fearful stimuli. Rats of a genetic strain with low CRH levels treat fearful situations with aplomb,[629] while rats with high CRH levels show signs of anxiety. In turn, the two CRH receptor systems do not act the same.[630,631] The CRH-1 system mediates a rapid fight-or-flight response, while the CRH-2 system is slower, mediating coping responses during the recovery phase of stress.[632] While CRH-1 gene knockout mice display markedly reduced anxiety,[633] CRH-2-deficient mutant mice are hypersensitive to stress.[634]

Thus we compare two states of high arousal, fear and sex, which are CNS states with opposite valences. We approach a sexual attraction; we avoid fear and pain. The contrast between approach responses and avoidance responses is universal and is the most fundamental to animal and human behavior. Classical ethologists highlighted this elementary dichotomy among animal behaviors, and a moment's reflection shows that it is central to human behavioral responses as well.[635] What performance characteristics do we expect (Table 6.1)

Table 6.1. Performance characteristics of neural systems reflecting high states of arousal

	Sex	Fear
Speed of response	Slow	Fast
Criterion for response	Optimize choice	Ensure survival
Mode of recovery	Satisfied	Regulated
Hormones involved	E, P, T	G, NE
Genes involved	ER-α, ER-β OT, OT-R	GR, MR CRH, CRH-R1 & 2
Neural circuitry in forebrain	Hypothalamic; Medial amygdala	Thalamic & cortical; Lateral & central amygdala

of neural systems reflecting high states of arousal but with opposite directional responses? The motor topographies of the two sets of responses are mutually exclusive. Therefore, the two sets of underlying mechanisms (isomorphic with the behaviors they support) must also oppose each other.

Approach responses can be slow. Their directions can be determined by long series of environmental signals over a substantial period of time. They result from choices among stimuli, intended to optimize the eventual result. Failure to do them is not fatal. Turning off the system is not important.

In contrast, avoidance responses have to be very fast. Failure to do them immediately may be fatal. Massive genetic redundancy and systems redundancy, with an emphasis on fast functional results, must therefore be anticipated. Then, because CNS circuits cannot stay in emergency status indefinitely, there must be a well-regulated way to turn the system off. Different genes are centrally involved (see Table 6.1). Neural circuits involved in fear, also, can be sharply distinguished from a sex behavior circuit at regulatory levels in the forebrain, as well as from the mechanisms for social affiliation.[539] As reviewed by LeDoux,[367] fear responses depend on the amygdala. For cell groups in the amygdala, I hypothetically contrast fear mechanisms with affiliative/sexual mechanisms, as follows. Pheromonal information critical for social behaviors in mice comes through the vomeronasal and main olfactory systems to the medial nucleus. Stimuli related to fear are processed in the lateral nucleus, with outputs exiting from the central nucleus. Thus we would predict central nucleus versus medial nucleus warfare to determine which dominates amygdala outputs for fear versus sex, respectively. At any given mo-

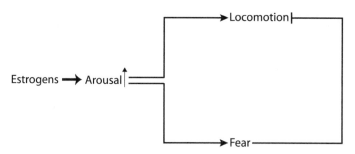

Figure 6.8. Environmental constraint on how estrogenically influenced arousal could affect specific behaviors. In a safe environment, increased arousal leads to increased locomotion, as would be appropriate for courtship behaviors. However, in a frightening environment, increased fear would have the opposite effect on locomotion. As envisioned by David Rubinow at NIMH (D. W. Pfaff et al., ed., *Hormones, Brain, and Behavior*, 2002), hormone/CNS/behavior relations in animals or humans depend upon context. (From M. Morgan et al., *Neurosci Biobehav Rev*, 2004; 28(1):55–63.)

ment in any given environment, the balance of power between these two systems, fear and sex, depends on both the species (e.g., prey vs. predator) and the individual's temperament.

If we know that sex hormones increase arousal, can we conclude that the hormone-loaded animal will always make approach responses, for example, locomotion in the service of courtship? No, the emotional sequelae of sex hormone–induced arousal may conflict with each other. They depend on environmental context. David Rubinow at the National Institutes of Health, speaking about human subjects,[611] and Maria Morgan at Rockefeller University, speaking about experimental animals,[636] have emphasized this dependency. In humans and animals, the individual behaviors that will be fostered by hormone treatment are determined by interactions with the individual's social and environmental surroundings (Fig. 6.8). For female mice in a safe environment, estrogen treatment causes increased motor activity that is appropriate for courtship behaviors. But in a novel environment or in contexts otherwise perceived as threatening, motor activity is instead reduced by estrogen treatment,[29] due to the hormone's arousing action that heightens fear. Note also that for explaining differences in human behavioral responses to estrogens, estrogen receptor (ER) genetic variants may have different effects on people's temperaments.[637]

In sum, the fundamental approach/avoidance dichotomy in human as well as animal behavior depends not only on hormone-stimulated arousal but also on the environmental context in which the individual makes the choice.

■ Applicability of Information Theory

For both fear and sex, the information content of the environmental situation influences the level of arousal. Let's examine the effects of unpredictability, irregularity, and lack of control on these two systems. Remember that in Chapter 1 these were essential elements of a high-information situation. For fear, anticipation of unknown bad stuff to come—angst—can be the worst part. For sex, anticipation of imagined delights may be the best part.

First, consider experimental studies with a pair of rats, one of which can control the delivery of a foot shock and the other of which cannot.[641] Which one gets more upset? The answer can be found in the propensity of the animals to get ulcers. The stomach lining of the "controlling" rat is clean, whereas the rat without control over its fate gets ulcers. This result supports my information theoretic approach. Unpredictability puts autonomic arousal over the top, with the result that arousal overload causes autonomic nervous changes that lead to ulcers. This theme plays out in a large number of stressful and fearful situations. Control over stressful elements in the environment reduces the harmful levels of arousal and upset, whether we are talking about the ability of mild stress to reduce pain,[638] to modulate the release of stress-related hormones from the pituitary gland,[639,640] to regulate arousal transmitters in the brain,[641,642] or to stimulate learning.[643,644] Mechanisms that carry this information theory effect include alterations of gene expression in the hypothalamus[645] and in the hippocampus.[646] Humans show this effect as well.[629] Unpredictability and uncontrollability of stress in the environment—for air traffic controllers, for example—are what cause the problem (Table 6.2). The requirement of constant vigilance to protect against sudden, unexpected, disastrous events will take its toll. In contrast, being informed of potentially painful events by a knowledgeable, trusted person reduces the harmful emotional information load and dissipates the worst of fear.

Table 6.2. Unpredictability and uncontrollability of stress in the environment

CONTROL		DEMAND	
		Low	High
	Low	restful	stressful
	High	boring	stimulating

This table illustrates a simple, classical idea recently spelled out in J. Rappaport and E. Seidman, *Handbook of Community Psychology* (New York: Kluwer Academic, 2000) and in J. Quick and L. Tetrick, *Handbook of Occupational Health Psychology* (Washington, D.C.: American Psychological Association, 2003).

For sexual arousal as well, informational content of the environment makes a difference. In terms of human psychology, we recognize the thrill of a bit of uncertainty about our potential partner. The anthropologist Helen Fisher has speculated that arousal-related transmitters participate in this phenomenon. Too much uncertainty, however, would be bad. Remember the inverted U curve in Figure 6.7. As well accepted as this idea might be for feelings of sexual arousal in humans, the information/sex connection gains its experimental footing in work with laboratory animals. Male rats that are sexually exhausted with one female undergo a sudden rejuvenation if placed with a new female. An unpredictable change makes all the difference. For female rats, constant mating is not a good idea. They prefer to receive a mount, wait for a while, and then get another mount by the male. Under these circumstances, and only these,[647] DA is released in anticipation of sexual behavior. For sex, anticipation of change with some lack of predictability, à la the equations in Chapter 1, maximizes the beneficial effects of arousal.

■ Libido and Stress in Humans

The elementary systems for producing heightened states of arousal are present in the human brain even as in animal brains. This is true both for positive emotions related to approach behaviors, like sexual desire, and for negative emotions, for example, those producing fear, anxiety, and stress. In addition, brain-scanning techniques such as functional magnetic resonance imaging (fMRI) have begun to reveal how emotional activity in the forebrain radiates out to the cerebral cortex in humans, the species with the most highly developed cerebral cortex.

Classically, scientists and behavioral economists who think about maximizing performance by humans (e.g., Kahneman[378]) did not talk much about arousal itself. Instead they talked about what is necessary to improve alertness and attention. In one of the most famous early models[378] (p. 18), brain arousal fuels the mental capacity necessary for attention and mental effort and is also heightened by muscular effort or even the emotions themselves. Human emotion, also, depends on arousal. In a theoretical tour de force, Robert Plutchik[648] of the Albert Einstein College of Medicine classified all emotions according to their intensity and their polarity (e.g., happy vs. sad, love vs. hate). Arousal accounts for the intensity of emotional experience. In fact, Plutchik[648] devised a special chart called a circumplex[648] (p. 71) in which the length of a vector described the intensity (the arousal level) implicit in an emotional display. Thus, respected theorists of human cognitive and emotional psychologies did make use of arousal-related concepts. What about mechanisms?

Because this chapter uses two heightened states of arousal with opposite emotional valences, sex and fear, to show how arousal mechanisms impact biologically important behaviors, let's begin with sex. It is almost embarrassing to see the extent to which the elementary mechanisms of sexual motivation have been preserved from fish to philosopher, from animal brain to human brain. This brief account rests on Chapter 8 in my book *Drive: Neural and Molecular Mechanisms for Sexual Motivation.*[20]

The chemistry of the sex steroid hormones, the chemistry and molecular biology of their receptor proteins, and the neuroanatomy of sex hormone receptor expression in the brain are essentially the same between animals and humans. Hypothalamic neuroanatomy, which is crucial for sex behavior, has remained much more similar between animals and humans than the neuroanatomy of other parts of the forebrain. Neurons with the highest concentrations of sex hormone receptors tend to project to other neurons with the same high concentrations of sex hormone receptors. Most exquisite is the ten-amino acid neuropeptide called gonadotropin-releasing hormone (GnRH, aka LHRH). It governs all of mammalian reproduction not only by controlling hormone release from the pituitary gland to the ovaries and testes but also, importantly, by facilitating mating behaviors.[548,649] Its chemistry, its molecular biology, and its neuroanatomy are preserved from animal to human brains. Striking is the developmental history of the GnRH neurons. They are the only neurons that migrate from outside the brain proper into the forebrain during embryonic life. Discovered in mice, this migration occurs in fish, in humans, and in every other vertebrate studied.

The similarities between animal and human brains are numerous. In terms of basic genetics and the molecular biology of neurons, the data so far indicate an essential identity of mechanisms across mammalian species. The fundamental chemistries of neuropeptides and neurotransmitters are also similar. Electrophysiologic signaling, synaptic mechanisms, and neuropharmacologic characteristics of cells in the human brain give us no reason to believe that these elements of sexual arousal mechanisms would be any different than in other mammals.

What about the reproductive biology that sexual arousal serves? From the elementary requirement for union between sperm and egg to the retention of sex differences in physiology and behavior, the similarities are more impressive than the differences as we compare human with lower mammals. Table 6.3 shows a partial list of primitive sex drive mechanisms preserved during evolution from animal to human brains.

What do these mechanisms buy us, as sexual beings? The founder of psychoanalysis, Sigmund Freud, who began his career as an M.D. and neurologist,

Table 6.3. Conservation of mechanisms from animals to human brain

Hormone Receptors	*Basic Biologic Mechanisms*
Ligands	Transcriptional controls
Receptor genes	Electrophysiology
Receptor chemistry	Synaptic physiology
Receptor neuroanatomy	Autonomic physiology
	Requirements for fertilization
CNS	*GnRH*
Hypothalamic neuroanatomy	Gene: coding region
Hypothalamic neuropeptides	Gene: promoter
Steroid receptors/neuronal connectivity	Developmental migration
	Adult neuroanatomy

invented the term "libido" to describe the urges and desires underlying the emotional, physical, and mental energies that result in sexual desire. He conceived of our libido as having two components—biologic and physiologic—and a complex psychological manifestation. Far be it from a modern neurobiologist to try to explain the psychological side—the full range of mental, artistic, self-conscious expressions of the person in love. However, we can claim to have the mechanisms in hand for the primitive, physiologic side of libido. Unless nature, having evolved a full set of working mechanisms for mammalian reproductive behavior, threw them all away and started an entirely new set for humans, we understand well, from mouse to madonna, the primitive mechanisms that drive sexual desire in humans. As a result, we have now gone well beyond psychoanalysis to unravel not only how a sex behavior works but also how sexual arousal comes about.

Mechanisms for negative emotional states—states that we prefer to avoid—also have been conserved as we move from animal brain to human brain.[186,650] Signals reflecting strong emotions modify higher cerebral processes associated with cognitive processes.[56] Neuroendocrine and neurophysiologic responses to frightening and alarming stimuli proceed in humans largely as they do in animals, and they are associated with an undesired emotional state usually called fear.[651] At least three concepts, separate but closely related, must be distinguished. *Fear* refers to our feeling that we want to avoid a specific thing. In most cases we feel fear for a limited amount of time. Neural circuitry for fear has been analyzed in animal brains.[367,627] *Anxiety*, especially free-floating anxiety, is much more generalized, may be hard to pin down to a specific fear, and may last a long time. A person can be generally anxious as part of his temperament.

Stress means a combination of reactions that add to each other: (1) the immediate physiologic response to a negative event, which includes both the hypothalamic and pituitary commands to the adrenal glands to release hormones like cortisol, plus the autonomic nervous system fight-or-flight response; and (2) the person's perception of both the negative event and his own physiologic response to it. Prolonged stress can make you sick because its hormonal and neural mechanisms affect the way in which your immune systems respond to pathogens.[629,652,653] The manner in which it does so differs between men and women.[654] Of course, these three concepts meld and can be combined. Repeated fearful incidents can cause multiple stress responses, for example, and foster a state of anxiety. Prolonged negative arousal states can foster depression[655] (also see Chap. 8). All of these concepts boil down to my *uncertainty about* the bad things that are going to happen to me. Novelty, unpredictability—essentially, the information content inherent in the dangers of the environment, present and future—are what feed these negative emotional states.

Brain-scanning techniques support our contention that, in the human CNS, those deep structures that manage heightened states of arousal associated with sex and fear remain in place. They radiate neuronal excitation to those cerebral cortical regions that represent the crowning glory of the human brain. The person imagining his or her beloved partner lights up his basal forebrain regions associated with dopamine projections, a neurotransmitter system supporting arousal and reward. In addition, a huge motor control region, the caudate, is activated[619,656] (cf. Arnow et al.[657]). Activations of certain parts of the midbrain and forebrain during orgasm have been shown by the Dutch neuroanatomist Gert Holstege to be basically the same for men and women.[658] Different brain regions glowed during magnetic resonance experiments conducted by the Emory University psychiatrist Helen Mayberg, who was scanning for activity when negative emotions such as fear or sadness were induced.

If these brain-scanning results show us how a specific arousal state can cause changes in widespread areas of brain, the reverse is also demonstrated in modern psychiatric medicine. Lack of arousal causes lack of sexuality. It is widely agreed that depressed people, especially women, hardly ever feel sexual desire and excitement. Both men and women have a much harder time reaching orgasm when depressed. Worse, some of the antidepressant medications, particularly selective serotonin reuptake inhibitors (SSRIs), have a paradoxical effect. They, too, reduce libido. This arousal-related set of problems is likely to receive much attention in pharmaceutical companies soon.

■ Summary

The order of discovery in the neurobiology of motivated behaviors began with the understanding of mechanisms for specific mating behaviors,[19,20] which depended on the discovery of estrogen receptors in individual neurons and the unraveling of the lordosis behavior circuit. It then proceeded to courtship behaviors involving locomotion and communication (approach responses), and now addresses the most fundamental issue: arousal itself. In terms of the functional genomics of the CNS, a set of modules, labeled GAPPS, show how effects of sex hormones on gene expression lead to sexually aroused behaviors. These highlight genetic networks operating through neural systems that are well understood, bringing the biologic analysis of mammalian behavior into the arena of modern functional genomics.

Once the brain is activated, what is a person to do? I contrast the performance characteristics required of systems underlying classic approach responses, as during sexual arousal, with systems responsible for classical avoidance responses, for example, during fear. In all of these heightened states of emotion, the informational content inherent in the situation determines our level of arousal. Information theory tells us that predictability and control bring down the level of arousal, which is good for fear but bad for sex.

7

Major Systems Questions about Brain Arousal Networks

Are the issues surrounding the fundamental force of the central nervous system sufficiently clear as to allow the kinds of questions electrical engineers and computer engineers ask of their circuits? If so, the conceptual tools of the physical sciences will be available to us as we understand changes of state in the brain.

Let's try to consider the gene networks and neural pathways underlying arousal in the way a design engineer would envision a new job in electronic circuit creation. Why? Because if the question-and-answer approach of this chapter is successful, some of the conceptual and mathematical techniques of control systems engineering may become available to us as neuroscientists.

Having all the intellectual tools possible to work on problems of CNS function is important. Thinking by analogy, facing elementary, primitive CNS arousal is like a geophysicist studying the magma at the core of the earth. In both cases, nothing is more fundamental to generating the body of facts at hand in the corresponding scientific field. We could draw another analogy to the field of astrophysics. Studying the very first arousal and alerting responses of the CNS is like researching the big bang at the beginning of our universe. In the two fields of science, before the first arousal or before the big bang, as far as that science is concerned, nothing happened.

During this exciting era of neuroscience, how do we conceive and phrase the most important theoretical questions? In the most abstract terms, can we discern the engineering design and structure of arousal mechanisms? Can we do the "mathematics of arousal"?

This chapter poses a number of questions about CNS arousal systems. I use some of the questions to organize the facts of the preceding chapters in a new and different way. Others are truly open questions intended to stimulate new thinking and experimentation.

▪ What Are Universal Operating Features of Arousal Systems?

What is the major, essential feature of arousal systems in the mammalian CNS? Answer: They are not allowed to fail. Therefore, overlapping functions among genes, neuropeptides, neurotransmitters, individual neurons, and nerve cell groups are required. The system's function must be protected against consequences from the loss of individual components. Redundancy is expected in neuroanatomic circuitry, neurophysiologic mechanisms, and especially among genes for receptors. Further, the high information content of arousal systems manifest in the low correlations among their components tends to protect against malfunction of individual elements.

Bilaterality. Are the systems symmetric, left for right? Yes, in every major respect I can think of, ascending arousal systems in the brainstem and descending controls—for example, from the paraventricular nucleus of the hypothalamus—are bilaterally symmetric. Even in circumstances in which connections look unbalanced left for right, eventual functional effects are closely coupled.[659] Across neurophysiologic research, however, there are exceptions. In the prefrontal cortex, high activity in the left compared to the right side favors joyful, even euphoric, states, while the reverse can lead to pathologic crying and depression.

Are both left and right sides of the neural circuits required? No. Unilateral damage can cause asymmetry in the EEG, specifically during REM sleep,[660] but it does not cause coma. Comatose states in patients occur only after serious bilateral damage to the systems specified in Chapter 2.

What are the implications of various forms of left/right connectivity? Do they add stability to arousal systems? Do they alter the frequency response of the system, or its response to repetitive stimulation? No one knows the answers to these questions.

Bipolarity (bidirectionality). Much of the neuroanatomy and neurophysiology reviewed in this book has highlighted how systems ascend from the brainstem and follow either a low road through the basal forebrain or a high road through the thalamus to affect the cerebral cortex. But are arousal systems essentially bipolar? Do they work in both directions at the same time? Yes. An important hypothalamic cell group, the paraventricular nucleus of the hypothalamus (PVN),[661,662] provides a wonderful example because it is involved in all four forms of arousal: cortical, autonomic, endocrine, and behavioral. Exactly how PVN neurons cause their wide range of effects remains a bit of a mystery. Projections to locus coeruleus could have something to do with PVN's alerting effects on the cortical EEG. And projections to autonomic control centers in the medulla and spinal cord[216,663] surely explain some of PVN's effects on autonomic arousal, through vasopressinergic, oxytocinergic, and corticotropin-releasing hormone (CRH) synapses. But we need much more neurochemical and biophysical detail to understand how PVN neurons can change the state of a mammalian brain and body.

Tucked within these long-distance connections from brainstem to basal forebrain or hypothalamus, and the reverse, is a much smaller set of reciprocal connections that comprise what Clifford Saper of Harvard Medical School has dubbed the sleep switch.[176] It appears that sleep-promoting neurons in the ventrolateral preoptic area not only receive ascending connections from wake-promoting histaminergic neurons in the tuberomammillary neurons of the hypothalamus but also send a descending projection back to those neurons.[166,174,175,664,665] He thinks that these connections provide a reciprocally opposed flip-flop control over the sleep-wake cycle. At issue is whether temperature controls in the preoptic area play an important causal role in governing waking and sleeping or whether they work simply as a parallel, correlated system.[323,666–669]

Another example of a descending pathway related to arousal would be that emanating from the brain's circadian clock, the suprachiasmatic nucleus (SCN) of the hypothalamus.[180] Taking in signals about daily changes in ambient light from the retina and feeding them to non-image-forming visual systems and elsewhere, the SCN both sends axons down the neuraxis and receives ascending signals. Later we'll face the question of what happens when rhythmic circadian signals encounter homeostatic sleep drive forces.

Preoptic area neurons, besides participating in Clifford Saper's sleep switch, have projections descending a longer distance that are related to autonomic nervous system controls as well as to temperature controls and sexual arousal.[116] Extensive descending connections control autonomic nervous system reactions, as determined from the neuroanatomic studies of Arthur Loewy and his colleagues at the Washington University School of Medicine[670] (reviewed, Loewy[671]; see also Luppi et al.[85] and Condes-Lara et al.[672]). In fact, I suspect that the reciprocal connections between brainstem arousal–related sites like the locus coeruleus and forebrain sites like the amygdala could provide the regulatory negative feedback required to bring rapid arousal changes under control.

Finally, we come to the master cells themselves. These large reticular neurons in the arousal crescent described in Chapter 2 (see Fig. 2.6) have descending as well as ascending axons.

In all of these examples, the bipolar, bidirectional nature of primitive arousal systems is clearly evident. Neural systems supporting arousal traverse the entire neural axis.[673] This is not to say, however, that all ascending and descending limbs are long-distance and monosynaptic. Some of the controls are ladder-like, with the essential connection being made in a series of shorter steps. How all of these components fit together is a question for advanced computer simulations using biologically realistic models and incorporating not only the connectivity reviewed in this book but also appropriate cellular and synaptic parameters.[674]

Are arousal systems polarized, back to front, or front to back? Does the brainstem arousal crescent largely bias the hypothalamic biologic clock, or is it the reverse? If the two levels of control cooperate in a dynamic fashion, then how do they do it? Regarding circadian rhythms of brain and behavior activation, can the lower brainstem arousal system influence frequency, phase, or amplitude? All of these questions are open for investigation.

BBURP. Trying to address the broadest questions we can possibly ask about CNS arousal systems, I propose a *bi*laterally symmetric, *bi*polar (bidirectional) *u*niversal *r*esponse *p*otentiating (BBURP) system. Among all vertebrates, this system readies the animal or the human to respond to stimuli of all modalities, to initiate voluntary locomotion, and to react with feeling to emotional challenges. Figure 7.1, which illustrates my BBURP theory, summarizes the most generalized features of arousal systems covered in this book.

Looking ahead, can we envision the mathematics of arousal? Yes. Following on from the neural, genetic, and neurochemical facts reviewed in Chapters 2–6, systematic trials with computer simulation will yield some of the insights. I applied Strogatz's theoretical proposition[234]—that certain highly connected elements dominate the emergence of order in complex systems (Chap. 2)—to our highly connected master cells (see Fig. 2.7). Also recall our discussion of Barabasi[228] using a power law to describe the numbers of connections achieved by elements in a population. I turned that into an idea about long-distance neural connections modulating activity in local modules (Fig. 2.7). Now, these ideas will be exciting to test in computer simulations and in real nervous systems.

Using both pictures and equations (Chap. 1), we can see that the totality of fundamental arousal is a function of generalized arousal plus an increasing function of several specific arousal states (sex, hunger, thirst, pain, fear, and so on) each with its own multiplying factor. To make further discoveries about CNS arousal systems, what is the best approach? Should we think of these neural pathways and chemical reaction systems as precisely designed in the

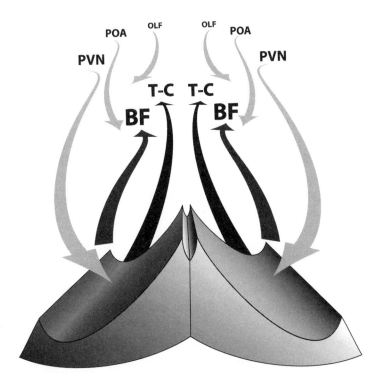

Figure 7.1. BBURP theory: A *b*ilaterally symmetric, *b*ipolar (bidirectional, ascending and descending) *u*niversal (among vertebrates) *r*esponse *p*otentiating system. This abstract, theoretical diagram is restricted to the major features of arousal systems that have been conserved throughout vertebrate phylogeny. Arising from an ancient, crescent-shaped field of neurons along the ventral and medial borders of the brainstem, arousing signals ascend. However, other important forces for regulating arousal descend from PVN, POA, and OLF. Abbreviations: BF, basal forebrain; OLF, olfactory and pheromonal inputs; POA, preoptic area; PVN, paraventricular nucleus of the hypothalamus; T-C, the nonspecific thalamocortical systems.

way that an electrical engineer or computer scientist would do things, or are they randomly and chaotically arrived at, with a nod toward evolution? In either case, can we consider them optimized in any respect—neuronal, hormonal, genetic—or can we see opportunities to improve arousal system performance in the case of challenge?

■ How Do We Meet the Requirement for Rapid Changes of CNS State?

Especially when confronted with dangerous circumstances, we must move into a state of high alertness and respond rapidly and adaptively. How? To answer this question theoretically, I suggest that positive feedback steps are required in arousal systems to achieve rapid transitions from soporific to alert states and to mount very fast and accurate responses. Similar to how an avalanche develops, ramping up certain arousal-related neurons facilitates other arousal-related neurons, resulting in a rapidly accelerating change in state.

I already see, in current neurobiologic evidence, many functional relations within arousal systems to support this idea. Not only do adrenergic inputs turn on basal forebrain cholinergic neurons and other adrenergic neurons in the locus coeruleus but also hypocretin neurons. In turn, hypocretin neurons excite other hypocretin neurons,[675] basal forebrain cholinergic neurons,[676] serotonergic neurons,[135] and adrenergic neurons in the locus coeruleus. For their part, serotonergic neurons receive not only hypocretin but also histaminergic and adrenergic inputs. Likewise, histamine neurons receive glutamatergic, hypocretin/orexin, cholinergic, and serotonergic inputs.[431] All of these data in this emerging field show us how activation of one segment of the arousal systems (for the neuroanatomy and genetics, see Chaps. 2 and 5) can hasten the activation of other segments. This produces a fast, cascading CNS response.

In neuroendocrine and autonomic neural systems as well, accelerating forces are evident. Valentino and her colleagues have shown how corticotropin-releasing hormone (CRH) raises the activity of locus coeruleus adrenergic neurons[104,284,677,678] (see also Conti and Foote[679]). Serotonin and noradrenaline activate vasopressin expression,[680] which in turn affects the cardiovascular and respiratory components of arousal.[663]

Given the data reviewed in Chapter 6, I suggest that these self-amplification mechanisms for producing changes in CNS state are not limited to generalized arousal. Ramping up generalized arousal enhances a specific arousal state, and vice versa. Think of the spring in your step after successful sexual encounters, and your heightened alertness and imagination when hungry.

Most of the time these "wind up" phenomena are good. In addition to the CNS being able to change state rapidly in response to circumstances, systems with a variety of time courses of internal regulatory feedback loops (fast as well as slow) have greater stability.[681,682] Other times, the positive feedback capacities of neuronal systems are a mixed blessing. Wind up phenomena in somatosensory pathways within the dorsal horn of the spinal cord foster chronic pain. Insofar as pain signals tissue damage to which we really must attend, this is good. But it is a pain.

If we begin to see how neurochemical and neurophysiologic arousal systems produce fast changes in CNS state, correspondingly we can reflect on the nature of their related behavioral phenomena. Emotions and arousal are quintessentially products of change, not of boring steady states. Therefore, the nature of the neural systems producing arousal and the nature of the behavioral requirements for those systems are congruent.

I have envisioned avalanche-like mechanisms to turn on generalized arousal systems quickly. Therefore I also must hypothesize mechanisms to turn the positive feedback *off*. Among the possibilities are linked negative feedback systems to regulate peak arousal levels, and exhaustion steps in which arousal-related neurons simply "run out of fuel." I do not know how this works. To explore a range of possible answers quickly, we have begun to simulate subsets of arousal mechanisms using methods pioneered by a generation of computer scientists, and thus we may be able to envision all the consequences, intended and unintended, of what I have proposed.

■ Sensitivity and Alacrity of Response, Yet Stability? How?

Arousal systems function within biologic entities. There are serious limitations on fuel supply, on the strength of the outputs, and on the range of quantitative values achievable in any physical dimension. Eventually, we must return to a resting state and achieve a new balance often described by the phrase "homeostatic system." This raises a question, as follows.

From the previous section of this chapter, there emerges a picture of arousal systems that rapidly accelerate changes in state. Yet, they must function in a balanced way, within defined limits. There is a creative tension between these two requirements. How can both be achieved? What are the critical neuronal parameters whose quantitative values shape the performance of the CNS? We do not know yet, and we are trying to simulate such systems to ask the what-if questions imaginatively and rapidly.

But for now, we can divide the theoretical answers into two parts, which are not mutually exclusive. First, intrinsic limits within each arousal-related neuron or within each arousal subsystem may limit the amplitude of response and enforce a return to baseline. Second, opposing actions among different genetic transcriptional systems related to arousal (Chap. 5) may, with varying time courses, dictate an overall balance by contradicting each other in their effects on behavioral outputs. For example, adrenergic, dopaminergic, and hypocretin/orexin systems can be limited in their arousal-fostering actions by prostaglandin D and opioid peptide–producing neurons (see references in Chap. 5). What about timing? When must synchrony of action be achieved among subsystems? Or when, instead, must precise time delays be achieved? These are open questions. Recognizing the constraints of any biologic system, I hypothesize that arousal mechanisms behave like a pendulum. They regain equilibrium by a dynamic in which, the farther they are from baseline state, the greater the force to return to baseline state. This theory can be tested by simulation, by behavioral studies, and by electrophysiologic experiments bearing on cortical arousal and autonomic arousal.

Any control system, biologic, electronic, or digital, faces problems like these.[683] Many of the questions I am raising for the CNS also find their analogies in immunology.[684] Mathematical theories of network dynamics are now seen as relevant to biologic questions,[685] for example, in studying the ways that network architecture and topology constrain system dynamics. I believe that the theoretical questions I am raising about primitive elementary arousal, Ur-Arousal, have answers that are manifest in a wide variety of physical and biologic systems. Therefore, thinking of arousal mechanisms as special examples of finite state automata may help us move forward.

Recognizing that some properties of arousal systems are extremely stable, we can ask, Do any of these neurochemical mechanisms apply to individual differences among humans in temperament? Genetically trained psychiatrists such as Robert Cloninger would give a resounding yes. Exactly how would long-term influences of differential gene expression impact responses to contemporary changes in the environment? We can answer these questions for some simple behaviors in mice,[686] but applications to human behavior are still lacking.

Does Automata Theory Apply?

Would it help to consider the brain as a special type of finite state automaton? Finite state automata are relatively simple computational devices characterized by their state diagrams.[108,687] They lend themselves to mathematical abstractions of systems concepts.[688] Perhaps thinking this way will allow us the conceptual tools of automata theory to explain arousal system performance. After all, as neurobiologists, we cannot use other tricks of computer science such as massive parallel processing because we have serious upper limits on our "hardware." But in terms of "software," our arousal mechanisms do seem to constitute information systems analogous to the types of software needed by chief executive officers to make decisions, and these feed motor pathways in the brain that themselves are analogous to the process automation systems needed by clerks.

Finite state automata are defined by their inputs, the descriptions of their states, the rules for transitions between states and their outputs.[689] Once the input alphabet, output alphabet, and state set have been specified, a given finite state automaton may be represented by its transition table—a display of the functions that govern changes in state upon certain inputs[690] (p. 17). The simplest have binary signals (go versus no-go) as their outputs. Others have outputs associated with a state (Moore machines) or with transitions between states (Mealy machines).

The neurobiologist is excited by the concept of circuit equivalence: two circuits are equivalent if we cannot tell them apart by their outputs no matter what the sequence of inputs.[691,692] Under what circumstances does the CNS perform very similar tasks with neural circuits that are far apart and superficially dissimilar, but that have a logical equivalence in Mealy's terms?

For applications of automaton theory to the CNS, it is interesting to think about simultaneous combinations of states representing different parts of the automaton. After all, it is probably too simplistic to think of the entire CNS in one state, especially given the "teamwork without identity" (Chap. 4) among different aspects of autonomic arousal, endocrine, cortical EEG, and behavioral arousal. The larger the number of possible combinations of states is, the larger the inherent information content of the automaton. Such issues make it necessary to do a certain amount of simulation to see the implications of this type of thinking.

One of the major themes in control systems engineering has to do with limitations and bounds within which input and output variables are allowed to swing (e.g., Jansson[693] and Kolda and Torczon[694]). These are hardest to figure out in multiple-input/multiple-output devices for which the relations and resonances among interior loops may not be anticipated. What are we optimizing: local system performance or intersystem overall performance? Accuracy or speed? In nonlinear control systems, low gain and low rates of change foster stability. In homeostatic systems even rapid, large-amplitude changes must allow critical parameters to return to baseline. This becomes a serious problem when feedback control systems are improperly designed and thus allow wild oscillations of their outputs. Applying this theme to the brain, our arousal systems must have been given protections against these wild oscillations, at least in the normal CNS. In the functional genomics of the CNS, for example, under what circumstances can single gene changes, affecting the complex interactions among gene products, throw a system out of bounds?[516]

Another issue has to do with what automaton states can lead to other specific states. What is the number of transitions required for any state A to lead to another state Z? In the CNS, surprising things happen. In Chapter 6 we asked whether the sign of a hormone effect on arousal might depend on the initial condition: is the animal or human starting out at the low end of the arousal continuum or at the high end? I was led to ask these questions because of having seen the following kind of electrical responses in neurons just forward of the hypothalamus, in the preoptic area.[695] In these neurons, if the animal's EEG arousal state was low and its resting activity was low, a specific olfactory stimulus would raise the rate of electrical discharges. However, if the EEG arousal state was higher and the animal's resting activity was high, the same olfactory stimulus would markedly decrease firing rates (Fig. 7.2). How does that work? I raise the question of whether *any* response of the CNS to *any* stimulus depends on the initial condition of the relevant cell group in the CNS.

Starting with the paper by McCulloch and Pitts,[696] neurobiologists have been eager to use the conceptual tools of computer science to understand the brain. Warren McCulloch[697] was intrigued by the reticular formation of the brainstem—included in my arousal crescent (Chap. 2)—as a behaviorally crucial iterative net with very low probabilities of neighboring neurons firing together and thus displaying high intrinsic information content. Under what circumstances can we optimize control system performance, either by altering quantitative values of parameters within our arousal systems to minimize errors or by altering dynamic response properties?[698,699] One neurobiologic problem to solve, evident in any biologic system, is how to get reliability from a device that has unreliable individual components.[700] In arousal pathways we know there is replication by redundancy of entire networks, which improves reliability. Second, insofar as some of our master cells in the arousal crescent are of a universal type, we probably have replication by repetition of certain elements within a network. Third, a circuit whose elements multiplex (multitask) offers the possibility of saving network function when individual modules go down. Fourth, we may be using error-correcting codes[700] within which redundancy, as calculated by Shannon's[701] type of math, specifies what is needed to preserve a given signal-to-noise ratio.

I note that certain designs of finite state automata have "trap states" from which emergence is not possible. Once a trap state has been entered, no combination of inputs, within the rules for transition in the automaton, allow it to be exited. Could this be analogous to coma, or to a permanent vegetative state? What about the possibility of recovery? In this context I am excited about the concept of decomposition of a finite state automaton. Even if certain automata are destroyed, their essential functions can be accomplished by two smaller automata neither of which is identical to the initial one but which achieve functional equivalence to that automaton.[687] Does this describe, in mathematical terms, part of what is achieved during recovery of the capacity for arousal following serious brain injury?

If this type of thinking opens theoretical opportunities for understanding the arousal systems of the brain, as I hope, some questions immediately arise.

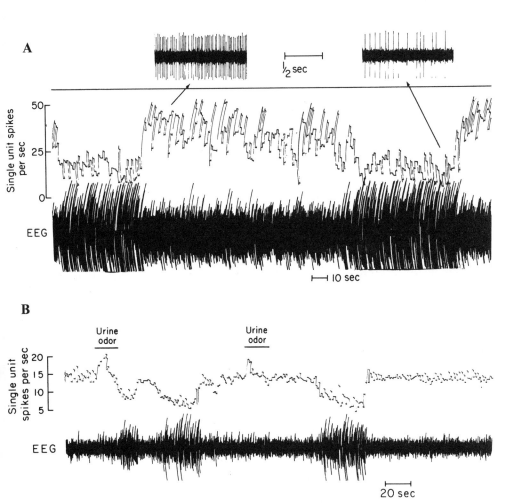

Figure 7.2. Opposite relations of single neuron activity to the cerebral cortical EEG, apparently related to the initial condition. Recordings from individual preoptic area neurons in urethane-anesthetized male rats. (A) Against a background of slow-wave high-amplitude EEG, the increase in neuronal activity is correlated with a switch to a high-frequency, low-amplitude EEG. (B) However, against a background of a high-frequency, low-amplitude EEG, presentation of a pheromone (female rat urine odor) can lead to an overall decrease in neuronal activity and change to a slow-wave high-amplitude EEG. (From Pfaff and Gregory, *Electroenceph Clin Neurophysiol*, 1971; 31: 223–30.)

First, the majority of serious treatments of control circuitry in electrical engineering and computer science have been with linear systems. Every neurobiologist would swear that we are dealing with exquisitely nonlinear systems. Our theories will need to treat advanced nonlinear multivariate control systems. What will be our best approaches?

Second, much of the history of control logic has been with clear and complete decision rules for generating crisply defined binary outputs following clear and discrete inputs. Is that the best way to think? What about fuzzy logic? Bart Kosko at the University of Southern California has reminded us in several books that the real world does not come at us in blacks and whites. It comes in many shades of grays. He insists that we think probabilistically. This insistence, of course, fits perfectly with my desire to bring information theory into neuroscience. Kosko's fuzzy principle states that if we are willing to give up the simplicity of bivalent yes/no thinking, we gain the accuracy of multivalent more/less thinking. Can we apply Kosko's approach to systems whose dynamic responses to ambiguous inputs are approximately this or that response, eventually, rounding into shape?

Finally, to aid us in thinking about modules within neural system arousal circuits, can we add the automaton concept of submachines[692] (pp. 73, 113)? A subautomaton B might be completely encompassed by the original automaton A but not have all the states and transitions that were defined by A. This type of conceptual apparatus might allow a formalization not only of functional subunits in the CNS but also an understanding of partial recoveries of function from diseases of the CNS.

■ Questions in the Time Domain

In all of the matters treated here, timing is an issue. I have claimed repeatedly that arousal is heightened by surprise, by change. The natural electrical engineering analogy is to a capacitor in an electronic circuit. A capacitor will not pass current when the circuit is in steady state. Only a sudden increase in voltage applied to the circuit will result in the capacitor allowing a peak of current to flow, which will then return exponentially to zero. Its mathematical expression is the first-order differential of voltage as a function of time, dv/dt.

What are the implications of differing time constants among arousal subsystems? If they vary widely among themselves, I would expect greater stability of performance. In particular, what are the effects of the very fastest functional connections? In the CNS, connexin protein expression forms gap junctions that offer the very rapid electrical signaling. Because these have been observed in locus coeruleus, we are pursuing the possibility that gap junction currents in these cells vastly improve the speed of system reaction as far as noradrenergic ascending pathways are concerned.

Chapter 4's discussion of autonomic arousal components included an analogy to a form of teamwork among arousal mechanisms—coordination *sans* correlation. Automata theory suggests some questions about this. Do arousal-related changes have to occur in a certain order? Is there a hierarchy of changes? As the CNS changes state extremely fast to react to an emergency, are there some frequencies of change at which arousal systems "resonate"? This question suggests that the frequency responses of all arousal components need investigation. To what extent do different aspects of brain arousal have to be synchronized? Or are there ideal delays among subsystems? In our own brains, are these parameters optimized, or is there room for improvement? Are alterations in timing a primary cause of dysfunction?

◼ Questions about Spatial Properties

We can ask similar questions about the effect of the topology of the arousal system. If we remember the neuroanatomy discussed in Chapter 2, it is fair to ask, Is it optimized? If not, how can we improve system performance either in the normal person or during recovery from insult? One issue is the possibility of modularity in arousal control systems. With respect to sexual arousal, when we unraveled the neural circuit for lordosis behavior, it was obvious from neuroanatomic and neurophysiologic data that there were at least five modules: spinal, hindbrain, midbrain, hypothalamic, and forebrain.[20] Luckily, these modules matched embryologic divisions of the CNS, and we could envision specific functions of each as they contributed to the regulation of lordosis behavior. What about our current concern for cortical arousal, neuroendocrine, autonomic, and behavioral arousal? Can modular controls be discerned for any of them?

Barabasi and his colleagues in the Department of Physics at Notre Dame[229,702] have changed our thinking about the spatial arrangements of biologic networks (Chap. 2). Rather than beginning with classical random networks, his theorizing assumes a power law in which a few highly interconnected nodes are linked by a relatively small number of long-distance links to less-connected nodes. His summary of the topological properties of metabolic networks in three diverse biologic domains turned out to be very similar to certain complex nonbiologic systems. How can this type of thinking help us during analyses of the CNS? Could my abstract idea of an arousal crescent (Chap. 2) represent one of his most highly connected nodes?

■ Thermodynamics, Information Theory, and Questions for the CNS

In his beautifully written book *Information, The New Language of Science*, Hans Christian von Baeyer[703] looks back on some informal approaches to conceiving of information subjectively, covers its modern definition by information theory, and looks forward to the possibility that information content will replace matter as the primary measurement in physical theory. Von Baeyer[703] talks about Richard Feynman's insight that information density links with the ability of science, physics in particular, to compress data (p. 16). Entire fields of common observations or experimental results can be represented accurately in a nifty, simple equation. Feynman's characterization leads to my own approach to the role of information theory in CNS function. To explain it, we have to take a detour through thermodynamics, the physical chemistry of energy, and heat. The Second Law of Thermodynamics states, as von Baeyer reminds us,[703] (p. 91) that without external intervention, entropy (heat divided by temperature) will remain constant or will increase over time. But what is it about heat divided by temperature? Ludwig Boltzmann, in Vienna, and John Willard Gibbs, in New Haven, the founders of thermodynamics, realized that heat divided by temperature refers to the number of ways that molecules in a given chemical entity, reaction, or space can be arranged[703] (p. 96). The greater the number of possibilities is, the less we know about their arrangement at time t. The concept of entropy, therefore, mapped perfectly onto Shannon's quantitative approach to the definition of information (Chap. 1).

It looks to me as though CNS arousal systems battle heroically against the Second Law of Thermodynamics in a very special way. They respond selectively to environmental situations that have an inherently high entropy—a high degree of uncertainty and therefore information content. But in responding, CNS arousal systems effectively reduce entropy by compressing all of that information into a single, lawful behavioral response.

Now, Shannon's mathematical definition of information content, derived from thermodynamics as a measure of uncertainty, is the bedrock of information theory. In biologic systems, the inevitable fluctuations of every measurable parameter make the essential information content higher, and influence the systems' responses to stimulation, whether studied in bacterial protein chemistry[704] or mammalian neurophysiology.[313] Entropic measurements make the ideas of information in the CNS more precise,[41] allowing us to quantify our hypotheses. They characterize nerve cell performance as these cells respond in nonlinear modes to stimuli that have several dimensions, presented in changing contexts in the presence of noise. In particular, they allow us to understand upper limits on system performance. Interval entropic measurements may constitute a further improvement.[705] Rieke and colleagues[41] note that the principles used in the application of informational thinking to the CNS are all independent of the exact stimuli or responses considered, or their meaning (p. 111). This approach is universally valid.

In human terms, the quantification of uncertainty allows us to understand the risks of gambling and the neurochemical response to the uncertainty of reward. These uncertainties command arousal, alertness, and attention correlated with higher firing rates by dopamine neurons. Importantly, the concepts of uncertainty and surprise, and thus information, have allowed Jerome Kagan, of Harvard University, to organize his thinking about clusters and patterns of human behavior in order to explain differences among individuals in temperament and personality.[1] What we must figure out now are the precise developmental mechanisms by which this well-documented set of personality differences—for example, inhibited vs. outgoing children—emerges.

Some of the major questions we will encounter while applying information theory to the arousal of the brain concern noise. Typically in neurophysiology, we have focused on maximizing the signal-to-noise ratio. However von Baeyer[703] points out at least three ways in which the presence of noise can be useful: (1) to amp up total signal power to reach the detection threshold of a receiver; (2) to permit "stochastic resonance" in which the true signal finds additional power by resonating with elements of noise; and (3) to provide "a thick blanket of sensory fog" that allows us to focus our attention on the most important stimuli in our environment (p. 126 ff). The neuroanatomy of Chapter 2 tells us that our brains clearly have the capacity to employ parallel arousal pathways to highlight an important stimulus and make a useful response. However, we still do not know exactly how this parallel processing works, and at the moment we can only state that the mammalian CNS cannot use "massive parallelism," as mentioned by von Baeyer[703] because of limitations on our brain's hardware (p. 194). Will computer-assisted brains of the future do better?

Finally, although I have trumpeted the pivotal role of Shannon's 1948 equation in crystallizing the concept of information, von Baeyer[703] raises the possibility that a new approach to information theory could supplant Shannon's (p. 217 ff). Indeed, this new approach by the Finnish-Swedish engineer Jan Kahre may be of special importance for neurobiologists. It adds a measure of reliability and, from game theory, a mathematical expression of the utility (value) of the information to the receiver. Thus, while I am certain that concepts of information are central to our understanding of arousal pathways in the brain, the best mathematical approach to the processing of salient signals of variable reliability and value, in the presence of noise, is still open to question.

◼ How Does a Sine Wave Impact a Sawtooth?

Two forces govern our changes in arousal during the twenty-four hours of the day-night cycle; and we have no idea, in mathematical or mechanistic terms, of how they interact. First, let's think of our underlying circadian rhythm as a sine wave (Fig. 7.3). The genetic bases of circadian clocks are increasingly well understood insofar as they involve transcriptional feedback loops.[184,185] They govern changes in behavior, in the autonomic nervous system,[706] and in the endocrine system.[180] Projections from SCN to the locus coeruleus must play some role.[90] One complexity to note is that the cyclic expression of clock genes certainly is not limited to the classical daily oscillator, the biologic clock generator in the suprachiasmatic nucleus (SCN) of the hypothalamus. Here are just three examples: cyclic expression is seen in the frontal cortex;[707] Garret Fitzgerald and his colleagues at the University of Pennsylvania have detected this expression in the cells of our vasculature; and adrenergic inputs impose a daily rhythm on the expression of clock genes *mperiod1 and mperiod 2* in liver cells.[708] How phased feedbacks from various organs and tissues throughout the body

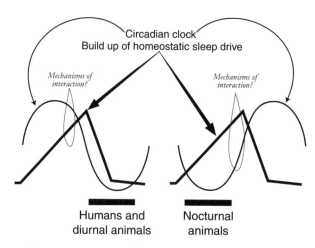

Figure 7.3. Circadian clock signals emanating from the suprachiasmatic nucleus of the hypothalamus must be integrated with fatigue building up from daily activities. The sleep neurophysiologist A. A. Borbely (*Hum Neurobiol*, 1982; 1: 195–204 and *Arch Ital Biol*, 2001; 139: 53–61) has faced this problem. The mechanisms through which these two major influences on arousal level interact are not yet understood.

onto the CNS would influence arousal during twenty-four hours is anyone's guess.

On the other hand, whatever the time of day, the longer we have been vigorous and active, the more tired we get. As A. A. Borbely has pointed out, peak arousal cannot last forever. We must crash. Let's think of the shape of this curve as an activity-driven and fatigue-limited sawtooth in time (see Fig. 7.3).

What happens functionally when the circadian sine wave meets the sawtooth-shaped sleep drive function? If they have the same phase and the same period of twenty-four hours, then there is no problem. The biologic clock's output from the SCN simply enforces our need to sleep. But what about all the alternative, more varied and subtle relations between sleep drive and circadian rhythms? Two experimental approaches have used pertubations of rest/activity cycles in humans[708a] and feeding times in mice.[708b] A deceptively easy sounding proposal is that the main oscillator in SCN drives the peripheral oscillators and that "the hypothalamus integrates them all,"[709] presumably including behavioral, EEG, autonomic, and hormonal rhythms. Such a proposal hides the complexity of interactions between sleep drives and circadian rhythms, because the devil lies in the details of the mechanisms therein. Further, we are just beginning to understand how the different outputs of the SCN relate to each other, even within the hypothalamus.[710] Here is a related question: How do powerful biochemical variables such as sex hormones, stress hormones, and thyroid hormones separately influence parameters of the clocks, features of sleep, and their interactions?

■ Unity from Diversity?

Throughout this book I have emphasized the importance of the diversity of CNS systems supporting arousal, from neuroanatomic pathways and electrophysiologic responses through genomic and neurochemical mechanisms. Within and between neuronal modules, low correlations of activities would make for a high-information throughput. In Chapter 4 we faced the lack of correlation among autonomic arousal measures and other indices of CNS arousal. Related questions have to do with specific arousal conditions, such as heightened states of fear and sex (Chap. 6). How do they interface with generalized arousal? Yes, after sex there may be that spring in the step (specific arousal feeding generalized arousal); and certainly a high state of generalized arousal will feed a person's fearfulness or sexiness. But how are essentially nonunitary networks and mechanisms transformed into an overall generalized arousal system? How does a multiplicity of mechanisms coalesce into a singularity of experience?

We have proven that such a function as generalized arousal exists (Chap. 1). Is it useful to struggle with the question of whether that generalized functional state is reified by a physical convergence of CNS arousal subsystems? Is that really necessary for a unified theory of arousal?

Most of the questions posed in this chapter have not yet found their answers. Some may not even be formulated very well. I present them with the hope that they will stimulate new work on arousal systems, both theoretical and experimental.

Summary and Practical Importance: From Biological Mechanisms to Health Applications

"But she might wake up," protested Susan . . . "Some of them wake up,"
shrugged Bellows, "but most don't."

Robin Cook, *Coma*, p. 42

▪ Main Points

During the twentieth century many neurobiologists considered their highest calling to understand the "particularity" of CNS responses. They asked, "Why would an animal make this particular behavioral response (but no other) to that specific stimulus (but no other)?" During the twenty-first century the emphasis will change. We will instead be trying to predict and explain changes in CNS states. We will examine the mechanisms that determine the conditions of entire brain systems. In turn, these systems modulate entire classes of responses to various sets of stimuli. This book addresses the most elementary question of CNS state: What determines the ability of an animal or person to mount any response to any stimulus, or to initiate voluntary motor activity, or to express an emotion? That is, I have reformulated the classic arousal problem.

Chapter 1 reviews the evidence that the CNS has a fundamental property called generalized arousal. The new operational definition of generalized arousal is: An animal or person who is more aroused *(1)* is more responsive to sensory stimuli; *(2)* emits more voluntary motor activity; and *(3)* is more emotionally reactive. A mathematical/statistical analysis of experimental results with mice showed that generalized arousal accounted for about one-third of arousal-related data. Thus, there *is* a set of mechanisms underlying generalized arousal, but there must be many specific forms of arousal as well, to account for the other two-thirds. At any given moment, our behavior is a compound function of a generalized arousal force supplemented by several specific arousal conditions. I have proposed an equation to express this conclusion in precise mathematical terms. If this equation expresses a lawfulness of arousal-related behavior, all of our handling of information and emotion must conform to it. At any rate, we now have the logical, concrete operational definition of generalized arousal, its mathematical expression, and a quantitative assay to make progress with this most fundamental property of the brain.

Arousal is the function that supports all cognitive abilities and all emotional expression. Cognition and emotion come together in arousal, which is why understanding this function constitutes a holy grail in neurobiology. Achieving this goal will bring together many strands of thinking by neuroscientists. Because we come from a history of ethologic thought (animal behavior and human ethology[12]) as well as from the experimental analysis of behavior, the elementary activation of brain and behavior was formerly seen to be an absolutely necessary concept but also a vague and slippery one. The arousal concept used to have the same undefined character as the concept of information did until Claude Shannon gave it a mathematical and experimental reality. Now, with an operational definition that will guide accurate experimental work, we can find out how arousal supports cognitive and emotional abilities. I suggest that it is through variations in arousal that cognition meets emotion.

Older theories tended to separate matters of thinking from matters of feeling. The evidence in this book shows that arousal mechanisms fuel both thinking and feeling. From the lower brainstem[711] through the ancient forebrain still present in humans[712] to the frontal cortex,[713–716] arousal theory helps us understand how cognitive and emotional capacities can be related.

Some of our habits of thinking and feeling may depend on genetic predispositions (Chap. 5) and may contribute to lifelong tendencies called cognitive and emotional styles.[3,334,717,718] Children who avoid novel or unfamiliar objects grow up into adults who have markedly different brain responses to novelty in their amygdala.[57] These arousal-related phenomena contribute to traits that have long been recognized in classical personality theory: enduring attitudes and tendencies typical of an individual.[2,719] Arousal mechanisms, involving autonomic arousal[721] as well as cortical and other neural mechanisms, contribute to our temperaments.[720] These involve autonomic arousal[721] as well as cortical and other neural mechanisms. Expressing our personal desires requires arousal and valence. This book explains the former.

Arousal neurobiology is the neuroscience of change, uncertainty, unpredictability, and surprise—that is, of information science. Throughout all of the analyses of arousal mechanisms in the CNS so far—neuroanatomic, physiologic, genetic, and behavioral—the concepts of information theory have proven useful. The mathematics of information provides ways of classifying neuronal responses to natural stimuli.[44,722] Nerve cells actually encode probabilities and uncertainties, with the result that they can guide behavior in unpredictable circumstances.[311] CNS arousal itself absolutely depends on change, uncertainty, unpredictability, and surprise. The huge phenomenon called habituation, a decline in response amplitude on repetition of the same stimulus, pervades neurophysiology, behavioral science, and autonomic physiology; and it shows us how declining information content leads to declining CNS arousal. Thus, arousal theory and information theory were made for each other.

In Chapter 2 we saw that a multiplicity of ascending arousal systems—different among themselves in their routes and their chemistries—prevent failure of this most basic CNS function. Because their activities cannot be highly correlated, their essential information content is high. Eventually, some of their axons converge. Those taking the high road affect thalamocortical arousal mechanisms. Those taking the low road work through the basal forebrain. In all cases I propose that relatively low numbers of long-distance connections control the activities of large numbers of neurons in "local modules."

Neurons generating the most primitive, powerful sources of arousal reside in a bilaterally symmetric crescent of neural tissue along the ventral and medial borders of the lower brainstem, stretching in a curved sheet from the medulla up to the midbrain. Within this arousal system are large reticular cells with unusually strong influence. These master cells have been highly conserved throughout vertebrate evolution, enduring certainly from the mouse brain and probably from the fish brain, to support arousal in the human brain. Some of them have both ascending and descending axons, a characteristic that emphasizes the bipolar, bidirectional nature of arousal systems.

In Chapter 3 we traveled up the brainstem charting neurons whose electrical responses are not specific to a particular stimulus modality, but instead are multimodal. They respond to virtually any stimulus that should and does alert the organism. Dopamine neurons comprise an especially interesting case in that they respond to salient stimuli, some of which are salient because they signal a special form of uncertainty: the probability of getting a reward. In these and other arousal systems, the information content of the stimulus can be related to response amplitude.

Olfactory inputs follow a different route to arousal stimulation. Because they project strongly to the amygdala, it is easy to understand how they influence behaviors occasioned by high arousal states such as sex and fear, especially in mammals that depend on olfaction. In this example we see how inputs signaling generalized arousal could influence behaviors that manifest specific arousal states (e.g., sex or fear).

In Chapter 4, which deals with autonomic nervous mechanisms related to arousal, I tossed aside the dogma that sympathetic and parasympathetic nervous systems are diametrically opposed to each other. Instead, I proposed that the two systems sometimes operate in tandem, in a temporally orchestrated manner. The key word is "patterns": the two systems' actions often achieve patterned teamwork, not simple opposition. Neurons in the hypothalamus help to generate and organize these temporal patterns of activity. Second, I attempted a reformulation of traditional thinking in the following manner: ascending arousal systems influencing the cerebral cortex were once thought to be completely separate from descending systems controlling autonomic outputs. Instead, I propose that sets of master neurons in the arousal crescent of the lower brainstem can influence both. The unity of the body is preserved.

The information content of autonomic reactions is high. Together with EEG arousal and actual behavioral arousal, these reactions display teamwork but not identity, coordination but not correlation. They also require high-information stimulus input; repeating the same stimuli leads to a decline of autonomic response. The efficient management of autonomic arousal supports hormone-driven as well as other emotional behaviors.

Chapter 5 describes genetic supports for the creation and maintenance of arousal systems. Their very multiplicity—more than 124 genes known so far, and counting—has at least four implications: (1) the number illustrates the importance of arousal functions; (2) by virtue of their different sites of expression and different chemistries, the genes offer protection against failure; (3) they provide for variation and flexibility of response; and (4) the genes intrinsically embody high-information content. Some increase arousal. Some decrease arousal. Recently cloned genes help to explain the behavioral effects of a common chemical such as caffeine and even the genesis of a rare malady, narcolepsy. Hormone-regulated gene expression shows how specific arousal forces can affect generalized arousal.

Chapter 6 addresses heightened states of arousal through two examples, sex and fear. These represent extremes among emotional states. They illustrate the deepest divide among behavioral responses: approach (for sex) versus avoidance (fear). Success in figuring out mechanisms for sex behavior emboldened us to propose genetic modules, networks that explain effects of sex hormones on courtship as well as mating (GAPPS modules). Underlying all sex and courtship behaviors is the concept of sexual motivation. The literature on motivation tells us that the expression of sexually motivated behavior requires the *activation* of behavior by a combination of sexual and generalized arousal.

Chapter 6 then contrasts sex with states of fear that require very fast avoidance responses. Both emotional states offer opportunities to discover the mathematical "trading relations" between generalized arousal and specific forms of arousal. We can study these using molecular and electrophysiologic as well as behavioral techniques. Both emotions feed on uncertainty, from the thrill of surprise during courtship to the depth of anxiety over anticipated pain. Throughout all of these investigations, it was apparent that the primitive CNS and hormonal mechanisms underlying sex and fear have been conserved from the animal to the human brain.

Chapter 7 pays tribute to our growing understanding of arousal neurobiology in two ways. First, I can now put together an overarching theory of primitive generalized arousal systems that are bilateral, bipolar, universal response potentiators (BBURP theory). I intend this to provide a standard model for elementary arousal mechanisms that support the generation of all behavioral responses in all vertebrates, including humans. Second, the field is becoming well enough developed that we can pose new questions based on systems theory and on the theory of finite state automata. We can ask such questions only if the CNS system under investigation has provided enough data with which to work.

Looking forward from a scientific point of view, arousal mechanisms will be exciting to analyze more deeply because they produce a universal, natural function of the CNS, a property of the brain extant throughout vertebrate evolution, from fish to philosopher. Now that we have a clear operational definition of generalized arousal, we discover that its mechanisms are easy to study. They are fast and thus occur first in the sequence of responses to any novel stimulus. In fact, they are causal to later responses. And they are important. The massive impacts of arousal malfunctions upon human activity are documented now.

■ Applications to Human Conditions

Problems involving the neuroscience of arousal, when totaled up, involve a large percentage of any nation's population. Ranging from purely medical problems such as comatose states and anesthesiology, through problems of mood, to public health problems such as lead poisoning, disorders of arousal demand a better understanding. The following sampling of topics gives an idea of the wide implications of problems in arousal systems that are fundamental to all cognitive and emotional functions.

When We Need to Elevate Arousal

Several medical conditions need increased CNS arousal for their amelioration.

States of reduced awareness. The most devastating medical problems related to arousal neurobiology are the vegetative, stuporous, and comatose states that follow many kinds of mechanical and vascular accidents to our heads.[252] These states frustrate doctors, who have not been able accurately to track all of the neurophysiologic parameters that might predict recovery in some cases, and they pose ethical dilemmas about the use of life support equipment. The first case that reached public attention in the United States was that of Karen Ann Quinlan in New Jersey, but more recent disputes among the family of Terry Schiavo in Florida have highlighted some of the legal issues. Hope is offered in two forms. First, equipped with new sophistication of thinking about all facets of arousal neurobiology, medical personnel can distinguish among patients with different degrees and types of loss of consciousness and treat them appropriately. For example, a definition has just been established for "minimally conscious state."[723–725] Second, our increasing understanding of ascending arousal systems may help neurologists and neurosurgeons to devise deep brain stimulation protocols that, combined with other forms of therapy, may be able to bring borderline patients to a higher state of awareness.

Here is a challenge: What can we do to understand the nerve cells and molecular mechanisms described in this book well enough to treat severe disorders of arousal? Can early intervention in patients with severe damage to this network rescue their cerebral life? Can later therapy in borderline vegetative patients return them to normal cognitive and emotional function?

A related set of tragic circumstances of great current importance follows from the widespread use of recreational drugs, especially by teenagers. Manipulating their arousal systems by any of a large number of compounds—ranging from historically notorious opium and LSD to those recently in vogue such as ecstasy and oxycontin[726]—these kids raise the possibility that they will live their lives with these crucial systems altered in their brains in ways we do not yet understand.[321] They have tampered with the balance normally achieved by the neuroanatomic and genetic systems covered in Chapters 2 and 5. It is a delicate equilibrium among complex systems designed to be labile precisely because of the important role of surprise and unpredictability in triggering rapid high-information responses to rapidly changing, high-information environments.

Problems of aging. These problems are not restricted to memory. Regarding autonomic arousal, during aging there is a declining ability of the sympathetic nervous system to respond to challenges, and a decrease in the plasticity of synaptic rearrangements in autonomic ganglia.[727] Depression in the aged may be as closely associated with autonomic/cardiovascular problems[728] as with specific changes in the forebrain. Clearly, declining arousal and alertness could contribute to whatever other problems of cognitive function or emotional expression might arise. Driving up brain arousal by exercise relieves some of the problems of the aged, and the neurochemical basis of that effect is under investigation.[729]

Fatigue states. The Centers for Disease Control (CDC) in Atlanta did not give credence to these conditions for a long time, but they affect millions of people in the United States and Europe. They all are defined by exclusion. Fibromyalgia syndrome (FMS) includes fatigue, generalized chronic pain, morning stiffness, sleep disturbance[730] and depression, and it occurs more frequently in women. Chronic fatigue syndrome (CFS) has many of the same symptoms, with less emphasis on pain and more emphasis on proinflammatory immune cell responses and increased reactivities to odors. It also is more frequent in young women. Both of these conditions are associated with low stress hormone responses, probably due to a blunted CRH response to stress in hypothalamic neurons.[731–734] Gulf War syndrome has symptoms almost identical to these and occurs more frequently in men. Both immune and neuroendocrine responses distinguish the fatigue states from primary depression.[735]

Alzheimer's. No one thinks that Alzheimer's disease is primarily an arousal disorder. However, it is highly unlikely to be a syndrome restricted to a simple loss of associative memory. Declines of elementary mental functions such as arousal are likely to be part of the cause. Any complex system is composed of many levels of operations. When a sophisticated output fails, a primitive subroutine may have been the cause. Consider an architectural analogy: If an old house has problems with windows on the third floor (higher functions), it may indeed be only a window problem. But settling of the foundation of the house (lower, primitive functions) may have contributed to the sticking of the windows on the third floor. Indeed, morphologic studies have revealed that Alzheimer's patients had cellular abnormalities in arousal-related neuronal groups[736] such as the locus coeruleus,[737, 738] the CRH neurons of the hypothalamus,[739] and the neurons in the hypothalamus that produce the arousal transmitter histamine.[740] Revving up arousal systems by exercise helps to ease some of the disease's effects. All of this reasoning convinces me that Alzheimer's patients should be tested for elementary cognitive and emotional functions, and that therapies that raise levels of autonomic, cortical, and behavioral arousal may help to delay onset of the most troubling stages of this terrible disease.

Autism. Autism is known as a failure of normal social recognition and response, but I feel that arousal-related reactions associated with meeting other people may be part of the problem. Cardiovascular reactivity constitutes an important component in our social behavior, and we react in particular, individual ways to social opportunities, challenges, and threats.[741] Thomas Boyce and his colleagues at the University of California at Berkeley have documented the large number of neuroendocrine and parasympathetic differences among children who occupy different positions in a social hierarchy. We all are socially context-sensitive, and part of that sensitivity is composed of our arousal reactions. When they are abnormal, we have a social problem.

Toxic states. What are the most sensitive and universally important ways to examine the harmful effects of toxins in the environment, especially those that might affect pregnant women and therefore growing baby brains? Lead in paint and in the water, mercury in fish, and heavy ingestion of alcohol by the mother all represent currently emphasized areas of concern. Effects on the development of the brain are most disturbing. Clearly, because tests of arousal can be made extremely sensitive and because elementary arousal is fundamental to all cognitive and emotional functions, an agency such as the Occupational Safety and Health Administration (OSHA) or the Environmental Protection Administration (EPA) must start here in order to claim a thorough set of measurements.

Problems of vigilance. Improving the duration and strength of arousal would be extremely important for the Department of Defense. The military has the problem of supporting soldiers', sailors', and pilots' performances and keeping them awake for long periods of time under extremely challenging conditions. These include high levels of stress and long periods of sleep deprivation. Likewise, OSHA must worry about workers in dangerous occupations such as in food-processing plants. Then there is the problem of shift work. A worker's ability to execute tasks efficiently and safely when shifted from a nighttime to an afternoon slot depends on the adequate performance of his or her arousal systems.

Arousal neurobiology and information theory, as presented in this book, set up a new science of wakefulness and effectiveness that should receive attention from pharmaceutical companies. We have tools such as amphetamine, methylphenidate, and modafinil and are beginning to understand their targets in the brain,[742] but we have a long way to go.

When We Need to Decrease Arousal

Counterintuitively, in some medical situations CNS arousal levels are higher than desired.

ADHD. Almost by definition an arousal disorder, attention deficit hyperactivity disorder (ADHD) puzzles me. I do not know why its frequency of diagnosis is rising, especially in the United States. Worse, we do not know why stimulants such as Ritalin work (when they work), and we are not sure whether children are being overmedicated. What is clear is ADHD's comorbidity with several other behavioral problems, aggressive and depressive, with overlaps that are different between boys and girls.

Sleep disorders. Insomnia, at any given time, affects about 15% to 35% of the adult population, depending on the criterion used (Partinen and Hublin, pp. 558–580, in Kryger, Roth, and Dement[743]). In every study, the incidence is greater in women than in men. The simplest definition of primary insomnia comes in three parts: The patient has trouble falling asleep, has trouble staying asleep, or wakes up very early. Primary insomnia sometimes overlaps with generalized anxiety disorder (Hauri, pp. 633–647, in Kryger, Roth, and Dement[743]).

The flip side of insomnia is unwanted daytime sleepiness, which happens commonly among 5% to 20% of the population, especially among sleep-deprived teenagers. It can be distinguished from frank narcolepsy, whose genetic basis is partly understood (see Chap. 5).

Sleep disorders are frequently comorbid with other arousal-related problems mentioned in this chapter. For example, they frequently are correlated with depression. And about 30% of children diagnosed with attention deficit disorder also have sleep problems.

Finally, sleep problems can be distinguished sharply from problems with circadian rhythms. In the latter case, we have simple shifts of phase or perhaps the period of the rhythm. In the former case, which is much more important, we have alterations of an underlying arousal-related variable, sleep drive, which normally builds up during wakefulness and declines as a result of the restorative properties of sleep.[744]

Sex offenders, violent aggression. Lust and rage are among the Bible's seven deadly sins. They show how inadequate handling of high generalized arousal can spill over into specific behavioral abnormalities. While hyperaggression in some families can be found to be associated with genetic changes[745] involving genes associated with monoamine transmitter regulation,[746] the overall problem of violence is much more complicated. Testosterone clearly is involved. However, violence among young men can be attributed to multiple causes, including socioeconomic disparities, problems during rearing, and psychological conditions.[747]

Hypersexuality, as in pedophilia, is a condition for which sex offenders are jailed. As a side point, consider the opposite problem: low sexual arousal, long noted among men and now recognized as a syndrome in women as well.

Post-traumatic stress disorder (PTSD). Sometimes it is important to be able to decrease arousal in order to reduce the harmful, lingering effects of a traumatic event. We need to know how to control certain high states of arousal (Chap. 6) in order to forget fear.

Hypertension. While it is obviously a problem of the autonomic nervous systems (Chap. 18 in Guyton and Hall[748]), high blood pressure also is frequently associated with a highly tense personality. Bjorntorp,[749] for example, portrays it as a disease that could originate with abnormalities in the hypothalamus. Hypertension is known to benefit from behavioral therapies such as relaxation techniques.

When Arousal Is Hard to Regulate

Fluctuations in CNS arousal are expected and necessary, but sometimes their proper regulation is difficult.

Mood disorders. In terms of mood, depression and manic-depressive (bipolar) disorders immediately come to mind. All of the neurobiologic systems that commonly are manipulated by efforts to raise or regulate a patient's mood were covered in Chapter 5, as we discussed the genetic and neurochemical bases of arousal. These systems are necessary for mood regulation. For example, transgenic mice that lack norepinephrine do not respond to antidepressant drugs.[750] Some of the mood-elevating properties of estrogenic hormonal therapies are laid to their actions on NE, DA, and 5HT systems. Again with regard to hormones, there are major differences between men and women in the incidence of depression, perhaps because of underlying gender differences in stress responses.[751–753] The arousal systems presented in this book are responsible for the activation of brain and behavior. As such, damage to them can reduce behavioral activation in a manner that we label "depression."

A recent trendy movie, *Coffee and Cigarettes*, is a quiet comedy composed of sketches in which couples manipulate their moods with these substances. How? The neuroanatomy covered in Chapter 2 coupled with the genetics and neurochemistry of Chapter 5 explain the roles of adenosine receptor blockers such as caffeine and nicotine derivatives such as cotinine acting through ACh receptors.

Throughout, I have argued that arousal systems depend on high-information content, unpredictability, and change. Consider that SSRIs act by clamping serotonin levels in synaptic clefts at high levels—high, but constant. This would explain the flat affect that is a troubling side effect of taking an SSRI for many patients.

Anesthesia. Anesthesiology is not a quantitative medical science. Depth of anesthesia is monitored by descriptive stages and is guided by probabilistic estimates of what a given alveolar concentration of a given agent does in a certain percentage of patients.[754] Autonomic nervous systems must maintain certain homeostatic balances during anesthesia.[755]

Major neurochemical systems available to be manipulated usefully to achieve anesthesia include GABA transmission and opioid peptide neuromodulation. Now, with transgenic mice we can begin to understand specific chemical mechanisms underlying general anesthesia. For example, changing a single asparagine residue in the β-2 subunit of the GABA-A receptor can change the sedation achieved by one general anesthetic, etomidate.[756] As far as hormones are concerned, some features of general anesthesia differ between men and women.[757]

Looking forward, I hope that adding the precision of a mathematical approach offered by information theory, using the depth of knowledge inherent in the neuroanatomy and electrophysiology reviewed in this book, and taking selective advantage of the new genetic information bearing on CNS arousal will help to make anesthesiology a safer and more solid branch of medical practice.

Alternative medicine. Does arousal neurobiology present a new approach to understanding the claims for medical benefits of meditation, relaxation techniques, religious belief, hypnosis, and so on? Only time will tell, but it does seem a fruitful avenue to explore.

In sum, the wide array of medical and public health problems to which this chapter speaks does not seem surprising, because arousal is fundamental to cognition and emotion. Therefore, its disorders would necessarily be multiple and serious. On the positive side, in normal behavior, changes in arousal are involved in emotional pleasures and rewards. For all of these reasons, genetic understanding of arousal mechanisms, pharmacological therapeutics, and behavioral/environmental modifications are all crucial to pursue as we search for solutions to disorders of arousal systems.

Works Cited

1. Kagan, J., *Surprise, uncertainty, and mental structures*. Cambridge, Mass.: Harvard University Press, 2002.
2. Eysenck, H. J., *The structure of human personality*, 3rd ed. London: Methuen, 1970.
3. Strelau, J., and H. J. Eysenck, *Personality dimensions and arousal: perspectives on individual differences*. New York: Plenum Press, 1987.
4. Hebb, D. O., *A textbook of psychology*. Philadelphia: Saunders, 1958.
5. Lorenz, K., *The foundations of ethology*. New York: Springer-Verlag, 1981.
6. Tinbergen, N., *The study of instinct*. Oxford: Clarendon Press, 1951.
7. VonHolst, E., and E. St.-Paul, *Vom Wirkungsgefuge der Triebe*. Naturwiss, 1960. 18: pp. 409–422.
8. Hinde, R., Interaction of internal and external factors in integrations of canary reproduction, in *Sex and behavior*, F. A. Beach, ed. New York: Wiley, 1965. pp. 381–415.
9. Hinde, R., *Animal behaviour: a synthesis of ethology and comparative psychology*. New York: McGraw-Hill, 1966.
10. Gallistel, C. R., *The organization of action: a new synthesis*. Hillsdale, N.J.: L. Erlbaum Associates, 1980; distributed by Halsted Press.
11. Immelmann, K., and C. Beer, *A dictionary of ethology*. Cambridge, Mass.: Harvard University Press, 1989.
12. Cofer, C. N., and M. H. Appley, *Motivation: theory and research*. New York: Wiley, 1964.
13. Hebb, D. O., Drives and the C.N.S. *Psychol Rev*, 1955. 62: pp. 243–254.
14. Hull, C. L., *A behavior system*. New Haven: Yale University Press, 1952.
15. Spence, K. W., *Behavior theory and conditioning*. New Haven: Yale University Press, 1956.
16. Duffy, E., *Activation and behavior*. New York: Wiley, 1962.
17. Haber, R. N., *Current research in motivation*. New York: Holt, 1966.
18. Bodnar, R., K. Commons, and D. Pfaff, *Central neural states relating sex and pain*. Baltimore: Johns Hopkins University Press, 2002.
19. Pfaff, D. W., *Estrogens and brain function: neural analysis of a hormone-controlled mammalian reproductive behavior*. New York: Springer-Verlag, 1980.
20. Pfaff, D. W., *Drive: neural and molecular mechanisms for sexual motivation*. Cambridge, Mass.: The MIT Press, 1999.
21. Moruzzi, G., and H. W. Magoun, Brain stem reticular formation and activation of the EEG. *J Neuropsychiatry Clin Neurosci*, 1949. 7(2): pp. 251–67.
22. Magoun, H. W., *The waking brain*, 2nd ed. Springfield: Charles C. Thomas, 1958.
23. Lindsley, D. B., L. H. Schreiner, W. B. Knowles, and H. W. Magoun, Behavioral and EEG changes following chronic brain stem lesions in the cat. *Electroenceph Clin Neurophysiol*, 1950. 2(4): pp. 483–98.
24. Lindsley, D. B., and D. W. Pfaff, Arousal, in *Encyclopedia of neuroscience*, G. Adelman, ed. Elsevier: Amsterdam, 2003.
25. Robbins, T., and B. Everitt, Arousal systems and attention, in *Handbook of cognitive neuroscience*, M. Gazzaniga, ed. Cambridge, Mass.: The MIT Press: 1996. pp. 703–20.
26. Garey, J., A. Goodwillie, J. Frohlich, M. Morgan, J. A. Gustafsson, O. Smithies, K. S.

Korach, S. Ogawa, and D. W. Pfaff, Genetic contributions to generalized arousal of brain and behavior. *Proc Natl Acad Sci U S A*, 2003. 100(19): pp. 11019–22.

27. Frohlich, J., M. Morgan, S. Ogawa, L. Burton, and D. Pfaff, Statistical analysis of hormonal influences on arousal measures in ovariectomized female mice. *Horm Behav*, 2002. 42(4): pp. 414–23.

28. Frohlich, J., Morgan, M., Ogawa, S., Burton, L. and Pfaff, D. W., Statistical analysis of measures of arousal in ovariectomized female mice. *Horm Behav*, 2001. 39: pp. 39–47.

29. Morgan, M. A., and D. W. Pfaff, Effects of estrogen on activity and fear-related behaviors in mice. *Horm Behav*, 2001. 40(4): pp. 472–82.

30. Morgan, M. A., and D. W. Pfaff, Estrogen's effects on activity, anxiety, and fear in two mouse strains. *Behav Brain Res*, 2002. 132(1): pp. 85–93.

31. Campbell, B. A., and F. D. Sheffield, Relation of random activity to food deprivation. *J Comp Physiol Psychol*, 1953. 46(5): pp. 320–2.

32. Barfield, R. J., and B. D. Sachs, Sexual behavior: stimulation by painful electrical shock to skin in male rats. *Science*, 1968. 161(839): pp. 392–3.

33. Antelman, S. M. and H. Szechtman, Tail pinch induces eating in sated rats which appears to depend on nigrostriatal dopamine. *Science*, 1975. 189(4204): pp. 731–3.

34. Brown, R. Z., Social behavior, reproduction and population changes in the house mouse (Mus musculus L.). *Ecological Monographs*, 1953. 23(3): pp. 217–40.

35. Richter, C., A behavioristic study of the activity of the rat. *Comp Psychology Monographs*, 1922. 1: pp. 1–55.

36. Deary, I. J., *Looking down on human intelligence: from psychometrics to the brain.* Oxford Psychology Series, no. 34. Oxford: Oxford University Press, 2000.

37. Gardner, H., *Frames of mind: the theory of multiple intelligences*, 10th anniversary ed. New York: Basic Books, 1993.

38. Shannon, C., A mathematical theory of communication. *Bell Sys Tech J*, 1948. 27: pp. 379–423.

39. Quastler, H., *Essays on the use of information theory in biology.* Urbana: University of Illinois Press, 1953.

40. Yockey, H. P., *Symposium on information theory in biology, Gatlinburg, Tennessee, October 29–31, 1956.* New York: Pergamon Press, 1958.

41. Rieke, F., et al., *Spikes: exploring the neural code.* Computational neuroscience. Cambridge, Mass.: The MIT Press, 1997.

42. Strong, S. P., R. R. de Ruyter van Steveninck, W. Bialek, and R. Koberle, On the application of information theory to neural spike trains. *Pac Symp Biocomput*, 1998: pp. 621–32.

43. Baddeley, R., P. J. B. Hancock, and P. Fèoldiâak, *Information theory and the brain.* Cambridge, England: Cambridge University Press, 2000.

44. Sharpee, T., N. C. Rust, and W. Bialek, Analyzing neural responses to natural signals: maximally informative dimensions. *Neural Comput*, 2004. 16(2): pp. 223–50.

45. Pfaff, D. W., Theoretical consideration of cross fiber pattern coding in the neural signaling of pheromones and other chemical stimuli. *Psychoneuroendocrinology*, 1975. 1: pp. 79–93.

46. Pagnoni, G., C. F. Zink, P. R. Montague, and G. S. Berns, Activity in human ventral striatum locked to errors of reward prediction. *Nat Neurosci*, 2002. 5(2): pp. 97–8.

47. Berns, G. S., S. M. McClure, G. Pagnoni, and P. R. Montague, Predictability modulates human *Brain Response* to reward. *J Neurosci*, 2001. 21(8): pp. 2793–8.

48. Bischoff-Grethe, A., M. Martin, H. Mao, and G. S. Berns, The context of uncertainty modulates the subcortical response to predictability. *J Cogn Neurosci*, 2001. 13(7): pp. 986–93.

49. Bischoff-Grethe, A., S. M. Proper, H. Mao, K. A. Daniels, and G. S. Berns, Conscious and unconscious processing of nonverbal predictability in Wernicke's area. *J Neurosci*, 2000. 20(5): pp. 1975–81.

50. Matsumoto, K., W. Suzuki, and K. Tanaka, Neuronal correlates of goal-based motor selection in the prefrontal cortex. *Science*, 2003. 301(5630): pp. 229–32.

51. Satoh, T., S. Nakai, T. Sato, and M. Kimura, Correlated coding of motivation and outcome of decision by dopamine neurons. *J Neurosci*, 2003. 23(30): pp. 9913–23.

52. Fiorillo, C. D., P. N. Tobler, and W. Schultz, Discrete coding of reward probability and uncertainty by dopamine neurons. *Science*, 2003. 299(5614): pp. 1898–902.

53. Komura, Y., R. Tamura, T. Uwano, H. Nishijo, K. Kaga, and T. Ono, Retrospective and prospective coding for predicted reward in the sensory thalamus. *Nature*, 2001. 412(6846): pp. 546–9.

54. Coop, A., M. Stavarache, D. W. Pfaff, and G. Reeke, Entropic analysis of behavior. Society for Neuroscience Abstracts, 2003.

55. Miller, G. A., The magic number seven, plus or minus two: some limits on our capacity for processing information. *Psychol Rev*, 1956. 64: pp. 81–97.

56. Panksepp, J., *Affective neuroscience: the foundations of human and animal emotions.* Series in Affective Science. New York: Oxford University Press, 1998.

57. Schwartz, C. E., C. I. Wright, L. M. Shin, J. Kagan, and S. L. Rauch, Inhibited and uninhibited infants "grown up": adult amygdalar response to novelty. *Science*, 2003. 300(5627): pp. 1952–3.

58. Kapp, B. and M. Cain, The neural basis of arousal, in *The international encyclopedia of social and behavioral sciences*, N. Smelser and P. Baltes, eds. Oxford: Elsevier Science, 2001. pp. 1463–66.

59. Asmus, S. E., and S. W. Newman, Tyrosine hydroxylase neurons in the male hamster chemosensory pathway contain androgen receptors and are influenced by gonadal hormones. *J Comp Neurol*, 1993. 331(4): pp. 445–57.

60. Sparks, D. L., and I. S. Nelson, Sensory and motor maps in the mammalian superior colliculus. *Trends Neurosci*, 1987. 10(8): pp. 312–317.

61. Wurtz, R. H., and J. E. Albano, Visual-motor function of the primate superior colliculus. *Annu Rev Neurosci*, 1980. 3: pp. 189–226.

62. Sherman, S. M., and R. W. Guillery, The role of the thalamus in the flow of information to the cortex. *Philos Trans R Soc Lond B Biol Sci*, 2002. 357(1428): pp. 1695–708.

63. Kim, D., D. Park, S. Choi, S. Lee, M. Sun, C. Kim, and H. S. Shin, Thalamic control of visceral nociception mediated by T-type Ca2+ channels. *Science*, 2003. 302(5642): pp. 117–9.

64. Desimone, R., and J. Duncan, Neural mechanisms of selective visual attention. *Annu Rev Neurosci*, 1995. 18: pp. 193–222.

65. Desimone, R., M. Wessinger, L. Thomas, and W. Schneider, Attentional control of visual perception: cortical and subcortical mechanisms. *Cold Spring Harb Symp Quant Biol*, 1990. 55: pp. 963–71.

66. Krout, K. E., R. E. Belzer, and A. D. Loewy, Brainstem projections to midline and intralaminar thalamic nuclei of the rat. *J Comp Neurol*, 2002. 448(1): pp. 53–101.

67. Ruggiero, D. A., S. Anwar, J. Kim, and S. B. Glickstein, Visceral afferent pathways to the thalamus and olfactory tubercle: behavioral implications. *Brain Res*, 1998. 799(1): pp. 159–71.

68. Zhang, D. X., and E. H. Bertram, Midline thalamic region: widespread excitatory input to the entorhinal cortex and amygdala. *J Neurosci*, 2002. 22(8): pp. 3277–84.

69. Mesulam, M. M., *Principles of behavioral neurology.* Contemporary Neurology Series, vol. 26. Philadelphia: F. A. Davis, 1985.

70. McCormick, D. A., and T. Bal, Sleep and arousal: thalamocortical mechanisms. *Annu Rev Neurosci*, 1997. 20: pp. 185–215.

71. Berridge, C. W., and B. D. Waterhouse, The locus coeruleus-noradrenergic system: modulation of behavioral state and state-dependent cognitive processes. *Brain Res Rev*, 2003. 42(1): pp. 33–84.

72. Foote, S. L., F. E. Bloom, and G. Aston-Jones, Nucleus locus coeruleus: new evidence of anatomical and physiological specificity. *Physiol Rev*, 1983. 63(3): pp. 844–914.

73. Amaral, D. G., and H. M. Sinnamon, The locus coeruleus: neurobiology of a central noradrenergic nucleus. *Prog Neurobiol*, 1977. 9(3): pp. 147–96.

74. Zardetto-Smith, A. M., and T. S. Gray, Organization of peptidergic and catecholaminergic efferents from the nucleus of the solitary tract to the rat amygdala. *Brain Res Bull*, 1990. 25: pp. 875–887.

75. Hokfelt, T., In vitro studies on central and peripheral monoamine neurons at the ultrastructural level. *Zeitschrift fur zellforschung*, 1968. 91: pp. 1–74.

76. Hokfelt, T., On the ultrastructural localization of noradrenaline in the central nervous system of the rat. *Zeitschrift fur Zellforschung*, 1967. 79: pp. 110–117.

77. Hokfelt, T., Distribution of noradrenaline storing particles in peripheral adrenergic neurons as revealed by electron microscopy. *Acta Physiol Scand*, 1969. 76: pp. 427–440.

78. Marrocco, R., and M. Davidson, Neurochemistry of attention, in *Mechanisms of attention*, R. Parasuraman, ed. Heidelberg: Springer, 1996. pp. 35–50.

79. Loughlin, S. E., S. L. Foote, and F. E. Bloom, Efferent projections of nucleus locus coeruleus: topographic organization of cells of origin demonstrated by three-dimensional reconstruction. *Neuroscience*, 1986. 18(2): pp. 291–306.

80. Berridge, C. W., and T. A. Stalnaker, Relationship between low-dose amphetamine-induced arousal and extracellular norepinephrine and dopamine levels within prefrontal cortex. *Synapse*, 2002. 46(3): pp. 140–9.

81. Berridge, C. W., T. L. Stratford, S. L. Foote, and A. E. Kelley, Distribution of dopamine beta-hydroxylase-like immunoreactive fibers within the shell subregion of the nucleus accumbens. *Synapse*, 1997. 27(3): pp. 230–41.

82. Rajkowski, J., P. Kubiak, S. Ivanova, and G. Aston-Jones, State-related activity, reactivity of locus coeruleus neurons in behaving monkeys. *Adv Pharmacol*, 1998. 42: pp. 740–4.

83. Schroeter, S., S. Apparsundaram, R. G. Wiley, L. H. Miner, S. R. Sesack, and R. D. Blakely, Immunolocalization of the cocaine- and antidepressant-sensitive l-norepinephrine transporter. *J Comp Neurol*, 2000. 420(2): pp. 211–32.

84. Zaborszky, L., and A. Duque, Sleep-wake mechanisms and basal forebrain circuitry. *Front Biosci*, 2003. 8: pp. 1146–69.

85. Luppi, P. H., G. Aston-Jones, H. Akaoka, G. Chouvet, and M. Jouvet, Afferent projections to the rat locus coeruleus demonstrated by retrograde and anterograde tracing with cholera-toxin B subunit and Phaseolus vulgaris leucoagglutinin. *Neuroscience*, 1995. 65(1): pp. 119–60.

86. Van Bockstaele, E. J., A. Saunders, P. Telegan, and M. E. Page, Localization of mu-opioid receptors to locus coeruleus-projecting neurons in the rostral medulla: morphological substrates and synaptic organization. *Synapse*, 1999. 34(2): pp. 154–67.

87. Van Bockstaele, E. J., J. Peoples, and P. Telegan, Efferent projections of the nucleus of the solitary tract to peri-locus coeruleus dendrites in rat brain: evidence for a monosynaptic pathway. *J Comp Neurol*, 1999. 412(3): pp. 410–28.

88. Van Bockstaele, E. J., D. Bajic, H. Proudfit, and R. J. Valentino, Topographic architecture of stress-related pathways targeting the noradrenergic locus coeruleus. *Physiol Behav*, 2001. 73(3): pp. 273–83.

89. Singewald, N., and A. Philippu, Release of neurotransmitters in the locus coeruleus. *Prog Neurobiol*, 1998. 56(2): pp. 237–67.

90. Aston-Jones, G., S. Chen, Y. Zhu, and M. L. Oshinsky, A neural circuit for circadian regulation of arousal. *Nat Neurosci*, 2001. 4(7): pp. 732–8.

91. Kow, L. M., G. D. Weesner, and D. W. Pfaff, α_1-adrenergic agonists act on the ventromedial hypothalamus to cause neuronal excitation and lordosis facilitation: electrophysiological and behavioral evidence. *Brain Res*, 1992. 588(2): pp. 237–45.

92. Kow, L.-M. and D. W. Pfaff, Estrogen effects on neuronal responsiveness to electrical and neurotransmitter stimulation: an in vitro study on the ventromedial nucleus of the hypothalamus. *Brain Res*, 1985. 347(1): pp. 1–10.

93. Fraley, G. S., Immunolesion of hindbrain catecholaminergic projections to the medial hypothalamus attenuates penile reflexive erections and alters hypothalamic peptide mRNA. *J Neuroendocrinol*, 2002. 14(5): pp. 345–8.

94. Cameron, N., P. Carey, and M. Erskine, Noradrenergic cells project directly to this medial amygdala. *Horm Behav*, 2003. 41: p. 459.

95. Rawson, J. A., C. J. Scott, A. Pereira, A. Jakubowska, and I. J. Clarke, Noradrenergic projections from the A1 field to the preoptic area in the brain of the ewe and Fos responses to oestrogen in the A1 cells. *J Neuroendocrinol*, 2001. 13(2): pp. 129–38.

96. Zhu, L., and T. Onaka, Involvement of medullary A2 noradrenergic neurons in the activation of oxytocin neurons after conditioned fear stimuli. *Eur J Neurosci*, 2002. 16(11): pp. 2186–98.

97. Buller, K., Y. Xu, C. Dayas, and T. Day, Dorsal and ventral medullary catecholamine cell groups contribute differentially to systemic interleukin-1-beta-induced hypothalamic pituitary adrenal axis responses. *Neuroendocrinology*, 2001. 73(2): pp. 129–38.

98. Berridge, C. W., A. F. Arnsten, and S. L. Foote, Noradrenergic modulation of cognitive function: clinical implications of anatomical, electrophysiological and behavioral studies in animal models. *Psychol Med*, 1993. 23(3): pp. 557–64.

99. Aston-Jones, G., M. Ennis, V. A. Pieribone, W. T. Nickell, and M. T. Shipley, The brain nucleus locus coeruleus: restricted afferent control of a broad efferent network. *Science*, 1986. 234(4777): pp. 734–7.

100. Aston-Jones, G., J. Rajkowski, P. Kubiak, R. J. Valentino, and M. T. Shipley, Role of the locus coeruleus in emotional activation. *Prog Brain Res*, 1996. 107: pp. 379–402.

101. Stone, E. A., G. L. Grunewald, Y. Lin, R. Ahsan, H. Rosengarten, H. K. Kramer, and D. Quartermain, Role of epinephrine stimulation of CNS alpha1-adrenoceptors in motor activity in mice. *Synapse*, 2003. 49(1): pp. 67–76.

102. Bouret, S., E. Kublik, and S. J. Sara, Rapid plasticity and learning-dependent interaction of medial frontal cortex and locus coeruleus neurons during go-no go odor discrimination in rat. Society for Neuroscience Abstracts, 2002: p. 284.11.

103. Pineda, J. A., S. L. Foote, and H. J. Neville, Effects of locus coeruleus lesions on auditory, long-latency, event-related potentials in monkey. *J Neurosci*, 1989. 9(1): pp. 81–93.

104. Valentino, R. J., S. L. Foote, and M. E. Page, The locus coeruleus as a site for integrating corticotropin-releasing factor and noradrenergic mediation of stress responses. *Ann N Y Acad Sci*, 1993. 697: pp. 173–88.

105. Fox, K., I. Wolff, A. Curtis, L. Pernar, E. J. Van Bockstaele, and R. J. Valentino, Multiple lines of evidence for the existence of corticotrophin-releasing factor (CRF) receptors on locus coeruleus (LC) neurons. Society for Neuroscience Abstracts, 2002: p. 637.9.

106. Curtis, A. L., and R. J. Valentino, Opioid receptor plasticity in locus coeruleus (LC) induced by swim stress: effects of intracoerulear damgo and hypotensive challenge. Society for Neuroscience Abstracts, 2002: p. 241.5.

107. Marson, L., and K. E. McKenna, CNS cell groups involved in the control of the ischiocavernosus and bulbospongiosus muscles: a transneuronal tracing study using pseudorabies virus. *J Comp Neurol*, 1996. 374(2): pp. 161–79.

108. Lewis, D. A., S. R. Sesack, A. I. Levey, and D. R. Rosenberg, Dopamine axons in primate prefrontal cortex: specificity of distribution, synaptic targets, and development. *Adv Pharmacol*, 1998. 42: pp. 703–6.

109. Carr, D. B. and S. R. Sesack, Dopamine terminals synapse on callosal projection neurons in the rat prefrontal cortex. *J Comp Neurol*, 2000. 425(2): pp. 275–83.

110. Carr, D. B. and S. R. Sesack, Terminals from the rat prefrontal cortex synapse on mesoaccumbens VTA neurons. *Ann N Y Acad Sci*, 1999. 877: pp. 676–8.

111. Isaac, S. O., and C. W. Berridge, Wake-promoting actions of dopamine D1 and D2 receptor stimulation. *J Pharmacol Exp Ther*, 2003. 307(1): pp. 386–94.

112. Giros, B., M. Jaber, S. R. Jones, R. M. Wightman, and M. G. Caron, Hyperlocomotion and indifference to cocaine and amphetamine in mice lacking the dopamine transporter. *Nature*, 1996. 379(6566): pp. 606–12.

113. Todtenkopf, M. S., T. Carreiras, R. H. Melloni, and J. R. Stellar, The dorsomedial shell of the nucleus accumbens facilitates cocaine-induced locomotor activity during the induction of behavioral sensitization. *Behav Brain Res*, 2002. 131(1–2): pp. 9–16.

114. Ralph, R. J., G. B. Varty, M. A. Kelly, Y. M. Wang, M. G. Caron, M. Rubinstein, D. K. Grandy, M. J. Low, and M. A. Geyer, The dopamine D2, but not D3 or D4, receptor subtype is essential for the disruption of prepulse inhibition produced by amphetamine in mice. *J Neurosci*, 1999. 19(11): pp. 4627–33.

115. Usiello, A., J. H. Baik, F. Rouge-Pont, R. Picetti, A. Dierich, M. LeMeur, P. V. Piazza, and E. Borrelli, Distinct functions of the two isoforms of dopamine D2 receptors. *Nature*, 2000. 408(6809): pp. 199–203.

116. Hull, E., R. Meisel, and B. D. Sachs, Male sexual behavior, in *Hormones, brain, and behavior*, D. W. Pfaff, A. Arnold, A.Etgen, S. Fahrbach, and R. Rubin, eds. San Diego: Academic Press/Elsevier, 2002. pp. 1–139.

117. Hull, E., J. Du, D. Lorrain, and L. Matuszewich, Testosterone, preoptic dopamine, and copulation in male rats. *Brain Res Bull*, 1997. 44: pp. 327–33.

118. Frye, C., and J. DeBold, 3α–OH-DHP and 5α–THDOC implants to the VTA facilitate sexual receptivity in hamsters after progesterone priming to the VMH. *Brain Res*, 1993. 612: pp. 130–137.

119. Frye, C., and E. Leadbetter, 5α-reduced progesterone metabolites are essential in hamster VTA for sexual receptivity. *Life Sci*, 1994. 54: pp. 653–59.

120. Becker, J. B., C. N. Rudick, et al., The role of dopamine in the nucleus accumbens and striatum during sexual behavior in the female rat. *J Neurosci*, 2001. 21(9): pp. 3236–41.

121. Dourish, C., and S. Iversen, Blockade of apomorhine induced yawning in rats by the dopamine autoreceptor antagonist (+)-AJ 76. *Neuropharmacology*, 1989. 28: pp. 1423–5.

122. Martin-Iverson, M., and S. Iversen, Day and night locomotor activity effects during administration of (+)-amphetamine. Pharmacology, Biochemistry and Behavior, 1989. 34: pp. 465–71.

123. Iversen, S., Behavioral topography in the striatum: differential effects of quinpirole and D-amphetamine microinjections. *Eur J Pharmacol*, 1998. 362: pp. 111–9.

124. Horvitz, J. C., Mesolimbocortical and nigrostriatal dopamine responses to salient nonreward events. *Neuroscience*, 2000. 96(4): pp. 651–6.

125. Robbins, T., S. Granon, J. Muir, F. Durantou, A. Harrison, and B. Everitt, Neural systems underlying arousal and attention: implications for drug abuse. *Ann N Y Acad Sci*, 1998. 846: pp. 222–237.

126. Schultz, W., P. Dayan, and P. Montague, A neural substrate of prediction and reward. *Science*, 1997. 275: pp. 1593–1599.

127. Wilson, C., G. G. Nomikos, M. Collu, and H. C. Fibiger, Dopaminergic correlates of motivated behavior: importance of drive. *J Neurosci*, 1995. 15(7 Pt 2): pp. 5169–78.

128. Ranaldi, R., D. Pocock, R. Zereik, and R. Wise, Dopamine fluctuations in the nucleus accumbens during maintenance, extinction, and reinstatement of intravenous D-amphetamine self-administration. *J Neurosci*, 1999. 19: pp. 4102–9.

129. Conrad, L., C. Leonard, and D. Pfaff, Connections of the median and dorsal raphe nuclei in the rat: an autoradiography and degeneration study. *J Comp Neurol*, 1974. 156: pp. 179–206.

130. Jacobs, B. L., and E. C. Azmitia, Structure and function of the brain serotonin system. *Physiol Rev*, 1992. 72(1): pp. 165–229.

131. Hermann, D. M., P. H. Luppi, C. Peyron, P. Hinckel, and M. Jouvet, Forebrain projections of the rostral nucleus raphe magnus shown by iontophoretic application of choleratoxin b in rats. *Neurosci Lett*, 1996. 216(3): pp. 151–4.

132. McCormick, D. A., Neurotransmitter actions in the thalamus and cerebral cortex and their role in neuromodulation of thalamocortical activity. *Prog Neurobiol*, 1992. 39(4): pp. 337–88.

133. Carr, D. B., D. C. Cooper, S. L. Ulrich, N. Spruston, and D. J. Surmeier, Serotonin receptor activation inhibits sodium current and dendritic excitability in prefrontal cortex via a protein kinase C–dependent mechanism. *J Neurosci*, 2002. 22(16): pp. 6846–55.

134. Peyron, C., P. H. Luppi, P. Fort, C. Rampon, and M. Jouvet, Lower brainstem catecholamine afferents to the rat dorsal raphe nucleus. *J Comp Neurol*, 1996. 364(3): pp. 402–13.

135. Brown, R. E., O. A. Sergeeva, K. S. Eriksson, and H. L. Haas, Convergent excitation of dorsal raphe serotonin neurons by multiple arousal systems (orexin/hypocretin, histamine and noradrenaline). *J Neurosci*, 2002. 22(20): pp. 8850–9.

136. el Kafi, B., R. Cespuglio, L. Leger, S. Marinesco, and M. Jouvet, Is the nucleus raphe dorsalis a target for the peptides possessing hypnogenic properties? *Brain Res*, 1994. 637(1–2): pp. 211–21.

137. Hermann, D. M., P. H. Luppi, C. Peyron, P. Hinckel, and M. Jouvet, Afferent projections to the rat nuclei raphe magnus, raphe pallidus and reticularis gigantocellularis pars alpha demonstrated by iontophoretic application of choleratoxin (subunit b). *J Chem Neuroanat*, 1997. 13(1): pp. 1–21.

138. Sun, X., J. Deng, T. Liu, and J. Borjigin, Circadian 5-HT production regulated by adrenergic signaling. *Proc Natl Acad Sci U S A*, 2002. 99(7): pp. 4686–91.

139. Jones, B. E., and A. C. Cuello, Afferents to the basal forebrain cholinergic cell area from pontomesencephalic—catecholamine, serotonin, and acetylcholine—neurons. *Neuroscience*, 1989. 31(1): pp. 37–61.

140. Cape, E. G., and B. E. Jones, Differential modulation of high-frequency gamma-electroencephalogram activity and sleep-wake state by noradrenaline and serotonin microinjections into the region of cholinergic basalis neurons. *J Neurosci*, 1998. 18(7): pp. 2653–66.

141. Fort, P., A. Khateb, A. Pegna, M. Muhlethaler, and B. Jones, Noradrenergic modulation of cholinergic nucleus basalis neurons demonstrated by in vitro pharmacological and immunohistochemical evidence in the guinea-pig brain. *Eur J Neurosci*, 1995. 7: pp. 1502–11.

142. Dringenberg, H., and C. Vanderwolf, 5-hydroxytryptamine (5-HT) agonists: effects on neocortical slow wave activity after combined muscarinic and serotonergic blockade. *Brain Res*, 1996. 728: pp. 181–7.

143. Vanderwolf, C., L. Leung, G. Baker, and D. Stewart, The role of serotonin in the control of cerebral activity: studies with intracerebral 5,7-dihydroxytryptamine. *Brain Res*, 1989. 504: pp. 181–91.

144. Watson, N., E. Hargreaves, D. Penava, L. Eckel, and C. Vanderwolf, Serotonin-dependent cerebral activation: effects of methiothepin and other serotonergic antagonists. *Brain Res*, 1992. 597: pp. 16–23.

145. Dringenberg, H., and C. Vanderwolf, Neocortical activation: modulation by multiple pathways acting on central cholinergic and serotonergic systems. *Exp Brain Res*, 1997. 116: pp. 160–174.

146. Boutrel, B., B. Franc, R. Hen, M. Hamon, and J. Adrien, Key role of 5-HT$_{1B}$ receptors in the regulation of paradoxical sleep as evidenced in 5-HT$_{1B}$ knock-out mice. *J Neurosci*, 1999. 19: pp. 3204–12.

147. Dringenberg, H. C., and C. H. Vanderwolf, Involvement of direct and indirect pathways in electrocorticographic activation. *Neurosci Biobehav Rev*, 1998. 22(2): pp. 243–57.

148. Alves, S. E., B. S. McEwen, S. Hayashi, K. S. Korach, D. W. Pfaff, and S. Ogawa, Estrogen-regulated progestin receptors are found in the midbrain raphe but not hippocampus of estrogen receptor alpha (ER alpha) gene-disrupted mice. *J Comp Neurol*, 2000. 427(2): pp. 185–95.

149. Uphouse, L., L. Colon, A. Cox, M. Caldarola-Pastuszka, and A. Wolf, Effects of mianserin and ketanserin on lordosis behavior after systemic treatment or infusion into the ventromedial nucleus of the hypothalamus. *Brain Res*, 1996. 718: pp. 46–52.

150. Uphouse, L., M. Andrade, M. Caldarola-Pastuszka, and A. Jackson, 5-HT1A receptor antagonists and lordosis behavior. *Neuropharmacology*, 1996. 35: pp. 489–95.

151. Maswood, N., M. Caldarola-Pastuszka, and L. Uphouse, 5-HT3 receptors in the ventromedial hypothalamus and female sexual behavior. *Brain Res*, 1997. 769: pp. 13–20.

152. Trevino, A., A. Wolf, A. Jackson, T. Price, and L. Uphouse, Reduced efficacy of 8-OH-DPAT's inhibition of lordosis behavior by prior estrogen treatment. *Horm Behav*, 1999. 35: pp. 215–222.

153. Jones, B. E., Arousal systems. *Front Biosci*, 2003. 8: pp. 438–51.

154. Hajszan, T., and L. Zaborszky, Direct catecholaminergic-cholinergic interactions in the basal forebrain. III. Adrenergic innervation of choline acetyltransferase-containing neurons in the rat. *J Comp Neurol*, 2002. 449(2): pp. 141–57.

155. Detari, L., K. Semba, and D. D. Rasmusson, Responses of cortical EEG-related basal forebrain neurons to brainstem and sensory stimulation in urethane-anaesthetized rats. *Eur J Neurosci*, 1997. 9(6): pp. 1153–61.

156. Wainer, B., and M.-M. Mesulam, Ascending cholinergic pathways in the rat brain, in *Brain cholinergic systems*, M. Steriade and D. Biesold, eds. New York: Oxford University Press, 1990.

157. Steriade, M., and G. Buzsaki, Parallel activation of thalamic and basal forebrain cholinergic systems, in *Brain cholinergic systems*, M. Steriade and D. Biesold, eds. New York: Oxford University Press, 1990.

158. Jones, B. E., Neurotransmitter systems regulating sleep-wake states, in *Biological psychiatry*, H. D'Haenen, J. A. den Boer, and P. Willner, eds. New York: Wiley, 2002.

159. Levy, R. B., and C. Aoki, Alpha7 nicotinic acetylcholine receptors occur at postsynaptic densities of AMPA receptor-positive and -negative excitatory synapses in rat sensory cortex. *J Neurosci*, 2002. 22(12): pp. 5001–15.

160. Erisir, A., A. I. Levey, and C. Aoki, Muscarinic receptor M(2) in cat visual cortex: laminar distribution, relationship to gamma-aminobutyric acidergic neurons, and effect of cingulate lesions. *J Comp Neurol*, 2001. 441(2): pp. 168–85.

161. Acuna-Goycolea, C., J. L. Valds, and F. Torrealba, Sequential activation of ascending

activating system neurons after exposure to food. Society for Neuroscience Abstracts, 2003: p. 473.9.

162. Vanni-Mercier, G., K. Sakai, and M. Jouvet, [Specific neurons for wakefulness in the posterior hypothalamus in the cat]. C R Acad Sci III, 1984. 298(7): pp. 195–200.

163. Khateb, A., P. Fort, A. Pegna, B. Jones, and M. Muhlethaler, Cholinergic nucleus basalis neurons are excited by histamine in vitro. *Neuroscience*, 1995. 69: pp. 495–506.

164. Easton, A. and D. Pfaff, Effects of histamine-1 receptor blockade on arousal responses in mice: sex differences. Pharmacology, biochemistry and behavior, 2004, in press.

165. Uhlrich, D. J., K. A. Manning, and J. T. Xue, Effects of activation of the histaminergic tuberomammillary nucleus on visual responses of neurons in the dorsal lateral geniculate nucleus. *J Neurosci*, 2002. 22(3): pp. 1098–107.

166. Sherin, J. E., J. K. Elmquist, F. Torrealba, and C. B. Saper, Innervation of histaminergic tuberomammillary neurons by GABAergic and galaninergic neurons in the ventrolateral preoptic nucleus of the rat. *J Neurosci*, 1998. 18(12): pp. 4705–21.

167. Reiner, P. B., and A. Kamondi, Mechanisms of antihistamine-induced sedation in the human brain: H1 receptor activation reduces a background leakage potassium current. *Neuroscience*, 1994. 59(3): pp. 579–88.

168. Servos, P., K. Barke, L. Hough, and C. Vanderwolf, Histamine does not play an essential role in electrocortical activation during waking behavior. *Brain Res*, 1994. 636: pp. 98–102.

169. Jones, B. E., P. Bobillier, C. Pin, and M. Jouvet, The effect of lesions of catecholamine-containing neurons upon monoamine content of the brain and EEG and behavioral waking in the cat. *Brain Res*, 1973. 58(1): pp. 157–77.

170. Jones, B. E., S. T. Harper, and A. E. Halaris, Effects of locus coeruleus lesions upon cerebral monoamine content, sleep-wakefulness states and the response to amphetamine in the cat. *Brain Res*, 1977. 124(3): pp. 473–96.

171. Posner, M. and S. Petersen, The attention system of the human brain. *Annu Rev Neurosci*, 1990. 13: pp. 25–42.

172. Coull, J., Neural correlates of attention and arousal: insights from electrophysiology, functional neuroimaging and psychopharmacology. *Prog Neurobiol*, 1998. 55: pp. 343–61.

173. Posner, M., and G. DiGirolamo, Attention in cognitive neuroscience: an overview, in *The cognitive neurosciences*, M. S. Gazzaniga and E. Bizzi, eds. 1995, Cambridge, Mass.: The MIT Press. pp. 623–31.

174. Sherin, J. E., P. J. Shiromani, R. W. McCarley, and C. B. Saper, Activation of ventrolateral preoptic neurons during sleep. *Science*, 1996. 271(5246): pp. 216–9.

175. Lu, J., M. A. Greco, P. Shiromani, and C. B. Saper, Effect of lesions of the ventrolateral preoptic nucleus on NREM and REM sleep. *J Neurosci*, 2000. 20(10): pp. 3830–42.

176. Saper, C. B., T. C. Chou, and T. E. Scammell, The sleep switch: hypothalamic control of sleep and wakefulness. *Trends Neurosci*, 2001. 24(12): pp. 726–31.

177. Zucker, I., M. Boshes, and J. Dark, Suprachiasmatic nuclei influence circannual and circadian rhythms of ground squirrels. *Am J Physiol*, 1983. 244(4): pp. R472–80.

178. Bittman, E. L., B. D. Goldman, and I. Zucker, Testicular responses to melatonin are altered by lesions of the suprachiasmatic nuclei in golden hamsters. *Biol Reprod*, 1979. 21(3): pp. 647–56.

179. Morin, L. P., The circadian visual system. *Brain Res Rev*, 1994. 19(1): pp. 102–27.

180. Kriegsfeld, L. J., J. LaSauter, T. Hamada, and R. Silver, Circadian rhythms in the endocrine system, in *Hormones, brain, and behavior*, D. W. Pfaff, A. Arnold, A.Etgen, S. Fahrbach, and R. Rubin, eds. San Diego: Academic Press/Elsevier, 2002. pp. 33–93.

181. Hogenesch, J. B., S. Panda, S. Kay, and J. S. Takahashi, Circadian transcriptional out-

put in the SCN and liver of the mouse. *Novartis Found Symp*, 2003. 253: pp. 171–80; discussion 52–5, 102–9, 180–3 passim.

182. Moore, R. Y., J. C. Speh, and R. K. Leak, Suprachiasmatic nucleus organization. *Cell Tissue Res*, 2002. 309(1): pp. 89–98.

183. Moore, R. Y., E. A. Abrahamson, and A. Van Den Pol, The hypocretin neuron system: an arousal system in the human brain. *Arch Ital Biol*, 2001. 139(3): pp. 195–205.

184. Young, M. W., Big Ben rings in a lesson on biological clocks. *Neuron*, 2002. 36(6): pp. 1001–5.

185. Young, M. W., and S. A. Kay, Time zones: a comparative genetics of circadian clocks. *Nat Rev Genet*, 2001. 2(9): pp. 702–15.

186. Damasio, A. R., *The feeling of what happens : body and emotion in the making of consciousness*, 1st ed. 1999, New York: Harcourt Brace.

187. Jansen, A. S., M. W. Wessendorf, and A. D. Loewy, Transneuronal labeling of CNS neuropeptide and monoamine neurons after pseudorabies virus injections into the stellate ganglion. *Brain Res*, 1995. 683(1): pp. 1–24.

188. Loewy, A. D., L. Marson, D. Parkinson, M. A. Perry, and W. B. Sawyer, Descending noradrenergic pathways involved in the A5 depressor response. *Brain Res*, 1986. 386(1–2): pp. 313–24.

189. Haxhiu, M. A., and A. D. Loewy, Central connections of the motor and sensory vagal systems innervating the trachea. *J Auton Nerv Syst*, 1996. 57(1–2): pp. 49–56.

190. Femano, P. A., S. Schwartz-Giblin, and D. W. Pfaff, Brain stem reticular influences on lumbar axial muscle activity. I. Effective sites. *Am J Physiol*, 1984. 246: pp. R389–95.

191. Femano, P. A., S. Schwartz-Giblin, and D. W. Pfaff, Brain stem reticular influences on lumbar axial muscle activity. II. Temporal aspects. *Am J Physiol*, 1984. 46: pp. R396–R401.

192. Cottingham, S. L., P. A. Femano, and D. W. Pfaff, Vestibulospinal and reticulospinal interactions in the activation of back muscle EMG in the rat. *Exp Brain Res*, 1988. 73: pp. 198–208.

193. Cottingham, S. L., P. A. Femano, and D. W. Pfaff, Electrical stimulation of the midbrain central gray facilitates reticulospinal activation of axial muscle EMG. *Exp Neurol*, 1987. 97: pp. 704–724.

194. Modianos, D., and D. W. Pfaff, Brain stem and cerebellar lesions in female rats. II. Lordosis reflex. *Brain Res*, 1976. 106: pp. 47–56.

195. Loy, D. N., D. S. Magnuson, Y. P. Zhang, S. M. Onifer, M. D. Mills, Q. L. Cao, J. B. Darnall, L. C. Fajardo, D. A. Burke, and S. R. Whittemore, Functional redundancy of ventral spinal locomotor pathways. *J Neurosci*, 2002. 22(1): pp. 315–23.

196. Van Bockstaele, E. J., V. A. Pieribone, and G. Aston-Jones, Diverse afferents converge on the nucleus paragigantocellularis in the rat ventrolateral medulla: retrograde and anterograde tracing studies. *J Comp Neurol*, 1989. 290(4): pp. 561–84.

197. Baffi, J. S., and M. Palkovits, Fine topography of brain areas activated by cold stress. A fos immunohistochemical study in rats. *Neuroendocrinology*, 2000. 72(2): pp. 102–13.

198. McKenna, K. E., Central nervous system pathways involved in the control of penile erection. *Annu Rev Sex Res*, 1999. 10: pp. 157–83.

199. Ladpli, R., and A. Brodal, Experimental studies of commissural and reticular formation projections from the vestibular nuclei in the cat. *Brain Res*, 1968. 8(1): pp. 65–96.

200. Willis, W. D., Jr., The pain system: the neural basis of nociceptive transmission in the mammalian nervous system. Pain Headache, 1985. 8: pp. 1–346.

201. Fort, P., P. H. Luppi, and M. Jouvet, Afferents to the nucleus reticularis parvicellularis of the cat medulla oblongata: a tract-tracing study with cholera toxin B subunit. *J Comp Neurol*, 1994. 342(4): pp. 603–18.

202. Wall, P. D., M. Lidierth, and P. Hillman, Brief and prolonged effects of Lissauer tract stimulation on dorsal horn cells. *Pain*, 1999. 83(3): pp. 579–89.

203. Mason, P., Contributions of the medullary raphe and ventromedial reticular region to pain modulation and other homeostatic functions. *Annu Rev Neurosci*, 2001. 24: pp. 737–77.

204. Curtis, K. S., E. G. Krause, and R. J. Contreras, Fos expression in non-catecholaminergic neurons in medullary and pontine nuclei after volume depletion induced by polyethylene glycol. *Brain Res*, 2002. 948(1–2): pp. 149–54.

205. Mayne, R. G., W. E. Armstrong, W. R. Crowley, and S. L. Bealer, Cytoarchitectonic analysis of Fos-immunoreactivity in brainstem neurones following visceral stimuli in conscious rats. *J Neuroendocrinol*, 1998. 10(11): pp. 839–47.

206. Hubscher, C. H., and R. D. Johnson, Inputs from spinal and vagal sources converge on individual medullary reticular formation neurons. Society for Neuroscience Abstracts, 2002: p. 271.2.

207. Lovick, T. A., The medullary raphe nuclei: a system for integration and gain control in autonomic and somatomotor responsiveness? *Exp Physiol*, 1997. 82(1): pp. 31–41.

208. Herbison, A. E., S. X. Simonian, N. R. Thanky, and R. J. Bicknell, Oestrogen modulation of noradrenaline neurotransmission. *Novartis Found Symp*, 2000. 230: pp. 74–85; discussion 85–93.

209. Bicknell, J., Estrogens addressing brainstem noradrenergic neurons with projections to hypothalamus, in *Stress and neuroendocrine systems*, H. Yamashita, ed. Heidelberg: Springer-Verlag, 1999.

210. Ramon-Moliner, E., An attempt at classifying nerve cells on the basis of their dendritic patterns. *J Comp Neurol*, 1962. 119: pp. 211–27.

211. Van Bockstaele, E. J., E. E. Colago, and S. Aicher, Light and electron microscopic evidence for topographic and monosynaptic projections from neurons in the ventral medulla to noradrenergic dendrites in the rat locus coeruleus. *Brain Res*, 1998. 784(1–2): pp. 123–38.

212. Johnson, A., G. Grunwald, J. Peoples, and E. J. Van Bockstaele, Evidence for opioid projections from the rostral ventral medulla to the nucleus of the solitary tract in rat brain. Society for Neuroscience Abstracts, 2002: p. 444.2.

213. Van Bockstaele, E. J., and G. Aston-Jones, Collateralized projections from neurons in the rostral medulla to the nucleus locus coeruleus, the nucleus of the solitary tract and the periaqueductal gray. *Neuroscience*, 1992. 49(3): pp. 653–68.

214. Aston-Jones, G., M. T. Shipley, G. Chouvet, Afferent regulation of locus coeruleus neurons: anatomy, physiology and pharmacology. *Prog Brain Res*, 1991. 88: pp. 47–75.

215. Farkas, E., A. S. Jansen, and A. D. Loewy, Periaqueductal gray matter input to cardiac-related sympathetic premotor neurons. *Brain Res*, 1998. 792(2): pp. 179–92.

216. Smith, J. E., A. S. Jansen, M. P. Gilbey, and A. D. Loewy, CNS cell groups projecting to sympathetic outflow of tail artery: neural circuits involved in heat loss in the rat. *Brain Res*, 1998. 786(1–2): pp. 153–64.

217. Farkas, E., A. S. Jansen, and A. D. Loewy, Periaqueductal gray matter projection to vagal preganglionic neurons and the nucleus tractus solitarius. *Brain Res*, 1997. 764(1–2): pp. 257–61.

218. Loewy, A. D., M. F. Franklin, and M. A. Haxhiu, CNS monoamine cell groups projecting to pancreatic vagal motor neurons: a transneuronal labeling study using pseudorabies virus. *Brain Res*, 1994. 638(1–2): pp. 248–60.

219. Strack, A. M., W. B. Sawyer, K. B. Platt, and A. D. Loewy, CNS cell groups regulating the sympathetic outflow to adrenal gland as revealed by transneuronal cell body labeling with pseudorabies virus. *Brain Res*, 1989. 491(2): pp. 274–96.

220. Manaker, S., and P. F. Fogarty, Raphespinal and reticulospinal neurons project to the dorsal vagal complex in the rat. *Exp Brain Res*, 1995. 106(1): pp. 79–92.
221. Schepens, B., and T. Drew, Postural role of the pontomedullary reticular formation during locomotion and reaching tasks in intact cats. Society for Neuroscience Abstracts, 2002: p. 854.2.
222. Dobbins, E. G., and J. L. Feldman, Brainstem network controlling descending drive to phrenic motoneurons in rat. *J Comp Neurol*, 1994. 347(1): pp. 64–86.
223. Chamberlin, N. L., and C. B. Saper, A brainstem network mediating apneic reflexes in the rat. *J Neurosci*, 1998. 18(15): pp. 6048–56.
224. Saper, C., Brain stem, reflexive behavior and cranial nerves, in *Principles of neural science*, E. R. Kandel, ed. New York: McGraw-Hill, 2000. pp. 874–888.
225. Stornetta, R. L., C. P. Sevigny, and P. G. Guyenet, Vesicular glutamate transporter DNPI/VGLUT2 mRNA is present in C1 and several other groups of brainstem catecholaminergic neurons. *J Comp Neurol*, 2002. 444(3): pp. 191–206.
226. Jones, B. E., Cytoarchitecture, transmitter and projections, in *The rat nervous system*, G. Paxinos, ed. New South Wales: Academic Press, 1995. pp. 155–71.
227. Manns, I. D., L. Mainville, and B. E. Jones, Evidence for glutamate, in addition to acetylcholine and GABA, neurotransmitter synthesis in basal forebrain neurons projecting to the entorhinal cortex. *Neuroscience*, 2001. 107(2): pp. 249–63.
228. Barabasi, A. L., *Linked: the new science of networks*. Cambridge, Mass.: Perseus, 2002.
229. Jeong, H., B. Tombor, R. Albert, Z. N. Oltvai, and A. L. Barabasi, The large-scale organization of metabolic networks. *Nature*, 2000. 407(6804): pp. 651–4.
230. Redner, S., How popular is your paper? An empirical study of the citation distribution. *Eur Phys J*, 1998. B4: pp. 131–4.
231. Ravasz, E., A. L. Somera, D. A. Mongru, Z. N. Oltvai, and A. L. Barabasi, Hierarchical organization of modularity in metabolic networks. *Science*, 2002. 297(5586): pp. 1551–5.
232. Barabasi, A. L., and R. Albert, Emergence of scaling in random networks. *Science*, 1999. 286(5439): pp. 509–12.
233. Albert, R., and A. L. Barabasi, Statistical mechanics of complex networks. *Reviews of Modern Physics*, 2002. 74: pp. 47–97.
234. Strogatz, S. H., *Sync: the emerging science of spontaneous order*, 1st ed. New York: Hyperion, 2003.
235. Watts, D. J., and S. H. Strogatz, Collective dynamics of "small-world" networks. *Nature*, 1998. 393(6684): pp. 440–2.
236. Nishizuka, M., and D. W. Pfaff, Medial preoptic islands in the rat brain: electron microscopic evidence for intrinsic synapses. *Exp Brain Res*, 1989. 77: pp. 295–301.
237. Nishizuka, M., and D. W. Pfaff, Intrinsic synapses in the ventromedial nucleus of the hypothalamus: an ultrastructural study. *J Comp Neurol*, 1989. 286: pp. 260–8.
238. Lee, R. S., S. C. Steffensen, and S. J. Henriksen, Discharge profiles of ventral tegmental area GABA neurons during movement, anesthesia, and the sleep-wake cycle. *J Neurosci*, 2001. 21(5): pp. 1757–66.
239. Jones, B., The organization of central cholinergic systems and their functional importance in sleep-waking states. *Prog Brain Res*, 1993. 98: pp. 61–71.
240. Steffensen, S. C., A. L. Svingos, V. M. Pickel, and S. J. Henriksen, Electrophysiological characterization of GABAergic neurons in the ventral tegmental area. *J Neurosci*, 1998. 18(19): pp. 8003–15.
241. Harrison, N. L., General anesthesia research: aroused from a deep sleep? *Nat Neurosci*, 2002. 5(10): pp. 928–9.
242. Tobler, I., C. Kopp, T. Deboer, and U. Rudolph, Diazepam-induced changes in sleep:

role of the alpha 1 GABA(A) receptor subtype. *Proc Natl Acad Sci U S A*, 2001. 98(11): pp. 6464–9.

243. Aoki, C., C. Venkatesan, C. G. Go, R. Forman, and H. Kurose, Cellular and subcellular sites for noradrenergic action in the monkey dorsolateral prefrontal cortex as revealed by the immunocytochemical localization of noradrenergic receptors and axons. *Cereb Cortex*, 1998. 8(3): pp. 269–77.

244. Kagan, J., *Galen's prophecy: temperament in human nature*. New York: Basic Books, 1994.

245. Thomas, A., and S. Chess, *Temperament and development*. New York: Brunner/Mazel, 1977.

246. Rothbart, M., and D. Derryberry, Development of individual differences in temperament, in *Advances in developmental psychology*, M. E. Lamb and A. L. Brown, eds. Hillsdale, N.J.: L. Erlbaum Associates, 1981. pp. 37–86.

247. Buss, A., and R. Plomin, A *temperament theory of personality development*. New York: Wiley, 1975.

248. Buss, A., and R. Plomin, T*emperament: early developing personality traits*. Hillsdale, N.J.: L. Erlbaum Associates, 1984.

249. Miller, R., ed. *Anesthesia*, 4th ed. New York: Churchill-Livingstone, 1994.

250. Nimmo, W., ed. *Anesthesia*. Oxford: Blackwell, 1994.

251. Plum, F., Coma and related generalized disturbances of the human conscious state, in *Cerebral cortex*, A. Peters, ed. New York: Plenum Publishing, 1991. pp. 359–425.

252. Plum, F., and J. B. Posner, *The diagnosis of stupor and coma*, 3rd ed. Philadelphia: Davis, 1982.

253. Plum, F., N. Schiff, U. Ribary, and R. Llinas, Coordinated expression in chronically unconscious persons. *Philos Trans R Soc Lond B Biol Sci*, 1998. 353: pp. 1929–1933.

254. Wheeler, R. E., R. J. Davidson, and A. J. Tomarken, Frontal brain asymmetry and emotional reactivity: a biological substrate of affective style. *Psychophysiology*, 1993. 30(1): pp. 82–9.

255. Edelman, G. M., and G. Tononi, *A universe of consciousness: how matter becomes imagination*, 1st ed. New York: Basic Books, 2000.

256. Parvizi, J., and A. Damasio, Consciousness and the brainstem. *Cognition*, 2001. 79(1–2): pp. 135–60.

257. Edelman, G. M., Neural darwinism: selection and reentrant signaling in higher brain function. *Neuron*, 1993. 10(2): pp. 115–25.

258. Lichtenberg, P., R. Bachner-Melman, I. Gritsenko, and R. P. Ebstein, Exploratory association study between catechol-O-methyltransferase (COMT) high/low enzyme activity polymorphism and hypnotizability. *Am J Med Genet*, 2000. 96(6): pp. 771–4.

259. Blood, A. J., and R. J. Zatorre, Intensely pleasurable responses to music correlate with activity in brain regions implicated in reward and emotion. *Proc Natl Acad Sci U S A*, 2001. 98(20): pp. 11818–23.

260. Leung, C. G., and P. Mason, Physiological survey of medullary raphe and magnocellular reticular neurons in the anesthetized rat. *J Neurophysiol*, 1998. 80(4): pp. 1630–46.

261. Leung, C. G., and P. Mason, Physiological properties of raphe magnus neurons during sleep and waking. *J Neurophysiol*, 1999. 81(2): pp. 584–95.

262. Li, Z., K. F. Morris, D. M. Baekey, R. Shannon, and B. G. Lindsey, Responses of simultaneously recorded respiratory-related medullary neurons to stimulation of multiple sensory modalities. *J Neurophysiol*, 1999. 82(1): pp. 176–87.

263. Li, Z., K. F. Morris, D. M. Baekey, R. Shannon, and B. G. Lindsey, Multimodal medullary neurons and correlational linkages of the respiratory network. *J Neurophysiol*, 1999. 82(1): pp. 188–201.

264. Potas, J. R., K. A. Keay, L. A. Henderson, and R. Bandler, Somatic and visceral afferents to the "vasodepressor region" of the caudal midline medulla in the rat. *Eur J Neurosci*, 2003. 17(6): pp. 1135–49.

265. Peterson, B. W., and C. Abzug, Properties of projections from vestibular nuclei to medial reticular formation in the cat. *J Neurophysiol*, 1975. 38(6): pp. 1421–35.

266. Peterson, B. W., M. E. Anderson, and M. Filion, Responses of ponto-medullary reticular neurons to cortical, tectal and cutaneous stimuli. *Exp Brain Res*, 1974. 21(1): pp. 19–44.

267. Evinger, C., C. R. Kaneko, and A. F. Fuchs, Activity of omnipause neurons in alert cats during saccadic eye movements and visual stimuli. *J Neurophysiol*, 1982. 47(5): pp. 827–44.

268. Pare, M., and D. Guitton, Brain stem omnipause neurons and the control of combined eye-head gaze saccades in the alert cat. *J Neurophysiol*, 1998. 79(6): pp. 3060–76.

269. Phillips, J. O., L. Ling, and A. F. Fuchs, Action of the brain stem saccade generator during horizontal gaze shifts. I. Discharge patterns of omnidirectional pause neurons. *J Neurophysiol*, 1999. 81(3): pp. 1284–95.

270. Drew, T., T. Cabana, and S. Rossignol, Responses of medullary reticulospinal neurones to stimulation of cutaneous limb nerves during locomotion in intact cats. *Exp Brain Res*, 1996. 111(2): pp. 153–68.

271. Lee, R. K., and R. C. Eaton, Identifiable reticulospinal neurons of the adult zebrafish, Brachydanio rerio. *J Comp Neurol*, 1991. 304(1): pp. 34–52.

272. Lee, R. K., R. C. Eaton, and S. J. Zottoli, Segmental arrangement of reticulospinal neurons in the goldfish hindbrain. *J Comp Neurol*, 1993. 329(4): pp. 539–56.

273. Heijdra, Y. F., and R. Nieuwenhuys, Topological analysis of the brainstem of the bowfin, Amia calva. *J Comp Neurol*, 1994. 339(1): pp. 12–26.

274. Eaton, R. C., R. K. Lee, and M. B. Foreman, The Mauthner cell and other identified neurons of the brainstem escape network of fish. *Prog Neurobiol*, 2001. 63(4): pp. 467–85.

275. Nakayama, H., and Y. Oda, Common sensory inputs and differential excitability of segmentally homologous reticulospinal neurons in the hindbrain. *J Neurosci*, 2004. 24(13): pp. 3199–209.

276. Casagrand, J. L., A. L. Guzik, and R. C. Eaton, Mauthner and reticulospinal responses to the onset of acoustic pressure and acceleration stimuli. *J Neurophysiol*, 1999. 82(3): pp. 1422–37.

277. Gahtan, E., N. Sankrithi, J. B. Campos, and D. M. O'Malley, Evidence for a widespread brain stem escape network in larval zebrafish. *J Neurophysiol*, 2002. 87(1): pp. 608–14.

278. Liu, K. S., and J. R. Fetcho, Laser ablations reveal functional relationships of segmental hindbrain neurons in zebrafish. *Neuron*, 1999. 23(2): pp. 325–35.

279. Rajkowski, J., H. Majczynski, E. Clayton, J. D. Cohen, and G. Aston-Jones, Phasic activation of monkey locus coeruleus (LC) neurons with recognition of motivationally relevant stimuli. Society for Neuroscience Abstracts, 2002: p. 86.10.

280. Clayton, E. C., J. Rajkowski, J. D. Cohen, and G. Aston-Jones, Activation of monkey locus coeruleus (LC) with stimulus identification in the Eriksen Flanker task. Society for Neuroscience Abstracts, 2002: p. 86.8.

281. Foote, S. L., G. Aston-Jones, and F. E. Bloom, Impulse activity of locus coeruleus neurons in awake rats and monkeys is a function of sensory stimulation and arousal. *Proc Natl Acad Sci U S A*, 1980. 77(5): pp. 3033–7.

282. Horvath, T. L., C. Peyron, S. Diano, A. Ivanov, G. Aston-Jones, T. S. Kilduff, and

A. N. van Den Pol, Hypocretin (orexin) activation and synaptic innervation of the locus coeruleus noradrenergic system. *J Comp Neurol*, 1999. 415(2): pp. 145–59.

283. Ivanov, A. and G. Aston-Jones, Hypocretin/orexin depolarizes and decreases potassium conductance in locus coeruleus neurons. *Neuroreport*, 2000. 11(8): pp. 1755–8.

284. Valentino, R. J., and S. L. Foote, Corticotropin-releasing hormone increases tonic but not sensory-evoked activity of noradrenergic locus coeruleus neurons in unanesthetized rats. *J Neurosci*, 1988. 8(3): pp. 1016–25.

285. Conti, L. H., K. L. Youngblood, M. P. Printz, and S. L. Foote, Locus coeruleus electrophysiological activity and responsivity to corticotropin-releasing factor in inbred hypertensive and normotensive rats. *Brain Res*, 1997. 774(1–2): pp. 27–34.

286. Ennis, M., and G. Aston-Jones, GABA-mediated inhibition of locus coeruleus from the dorsomedial rostral medulla. *J Neurosci*, 1989. 9(8): pp. 2973–81.

287. Gervasoni, D., L. Darracq, P. Fort, F. Souliere, G. Chouvet, and P. H. Luppi, Electrophysiological evidence that noradrenergic neurons of the rat locus coeruleus are tonically inhibited by GABA during sleep. *Eur J Neurosci*, 1998. 10(3): pp. 964–70.

288. Foote, S. L., C. W. Berridge, L. M. Adams, and J. A. Pineda, Electrophysiological evidence for the involvement of the locus coeruleus in alerting, orienting, and attending. *Prog Brain Res*, 1991. 88: pp. 521–32.

289. Horvitz, J. C., T. Stewart, and B. L. Jacobs, Burst activity of ventral tegmental dopamine neurons is elicited by sensory stimuli in the awake cat. *Brain Res*, 1997. 759(2): pp. 251–8.

290. Horvitz, J. C., Dopamine gating of glutamatergic sensorimotor and incentive motivational input signals to the striatum. *Behav Brain Res*, 2002. 137(1–2): pp. 65–74.

291. Paladini, C. A., S. Robinson, H. Morikawa, J. T. Williams, and R. D. Palmiter, Dopamine controls the firing pattern of dopamine neurons via a network feedback mechanism. *Proc Natl Acad Sci U S A*, 2003. 100(5): pp. 2866–71.

292. Le Moal, M. and H. Simon, Mesocorticolimbic dopaminergic network: functional and regulatory roles. *Physiol Rev*, 1991. 71: pp. 155–210.

293. Blackburn, J. R., J. G. Pfaust, and A. G. Phillips, Dopamine functions in appetitive and defensive behaviors. *Prog Neurobiol*, 1992. 39: pp. 247–279.

294. Jouvet, M., Etude de la dualite des etats de sommeil et des mecanismes de la phase paradoxale, in *Aspects anatomo-fonctionnels de la physiologie du sommeil*, M. Jouvet, ed. Paris: Centre Nat Rech Sci, 1965. pp. 393–442.

295. Jacobs, B. L., and C. A. Fornal, Activity of brain serotonergic neurons in the behaving animal. *Pharmacol Rev*, 1991. 43(4): pp. 563–78.

296. Fornal, C. A., W. J. Litto, D. A. Morilak, and B. L. Jacobs, Single-unit responses of serotonergic neurons to glucose and insulin administration in behaving cats. *Am J Physiol*, 1989. 257(6 Pt 2): pp. R1345–53.

297. Veasey, S. C., C. A. Fornal, C. W. Metzler, and B. L. Jacobs, Response of serotonergic caudal raphe neurons in relation to specific motor activities in freely moving cats. *J Neurosci*, 1995. 15(7 Pt 2): pp. 5346–59.

298. Veasey, S. C., C. A. Fornal, C. W. Metzler, and B. L. Jacobs, Single-unit responses of serotonergic dorsal raphe neurons to specific motor challenges in freely moving cats. *Neuroscience*, 1997. 79(1): pp. 161–9.

299. Fornal, C. A., W. J. Litto, C. W. Metzler, F. Marrosu, K. Tada, and B. L. Jacobs, Single-unit responses of serotonergic dorsal raphe neurons to 5-HT1A agonist and antagonist drug administration in behaving cats. *J Pharmacol Exp Ther*, 1994. 270(3): pp. 1345–58.

300. Jacobs, B. L., and C. A. Fornal, Activity of serotonergic neurons in behaving animals. *Neuropsychopharmacology*, 1999. 21(2 Suppl): pp. 9S-15S.

301. Jacobs, B. L. and C. A. Fornal, Serotonin and motor activity. *Curr Opin Neurobiol*, 1997. 7(6): pp. 820–5.

302. Manohar, S., H. Noda, and W. R. Adey, Behavior of mesencephalic reticular neurons in sleep and wakefulness. *Exp Neurol*, 1972. 34(1): pp. 140–57.

303. Kasamatsu, T., Maintained and evoked unit activity in the mesencephalic reticular formation of the freely behaving cat. *Exp Neurol*, 1970. 28(3): pp. 450–70.

304. Scheibel, M., A. Scheibel, A. Mollica, and G. Moruzzi, Convergence and interaction of afferent impulses on single units of reticular formation. *J Neurophysiol*, 1955. 18(4): pp. 309–31.

305. Amassian, V. E., and R. V. Devito, Unit activity in reticular formation and nearby structures. *J Neurophysiol*, 1954. 17(6): pp. 575–603.

306. Fanselow, E. E., K. Sameshima, L. A. Baccala, and M. A. Nicolelis, Thalamic bursting in rats during different awake behavioral states. *Proc Natl Acad Sci U S A*, 2001. 98(26): pp. 15330–5.

307. Hartings, J. A., S. Temereanca, and D. J. Simons, State-dependent processing of sensory stimuli by thalamic reticular neurons. *J Neurosci*, 2003. 23(12): pp. 5264–71.

308. Aggleton, J. P., *The amygdala: a functional analysis*, 2nd ed. Oxford: Oxford University Press, 2000.

309. Hamann, S., R. A. Herman, C. L. Nolan, and K. Wallen, Men and women differ in amygdala response to visual sexual stimuli. *Nat Neurosci*, 2004. 7(4): pp. 411–6.

310. Crabtree, J. W., Intrathalamic sensory connections mediated by the thalamic reticular nucleus. *Cell Mol Life Sci*, 1999. 56(7–8): pp. 683–700.

311. Glimcher, P. W., *Decisions, uncertainty, and the brain: the science of neuroeconomics*. Cambridge, Mass.: The MIT Press, 2003.

312. Schultz, W., Predictive reward signal of dopamine neurons. *J Neurophysiol*, 1998. 80(1): pp. 1–27.

313. Foldy, C., I. Aradi, A. Howard, and I. Soltesz, Diversity beyond variance: modulation of firing rates and network coherence by GABAergic subpopulations. *Eur J Neurosci*, 2004. 19(1): pp. 119–30.

314. Lindsley, D. B., J. W. Bowden, and H. W. Magoun, Effect upon the EEG of acute injury to the brain stem activating system. *Electroenceph Clin Neurophysiol*, 1949. 1: pp. 475–86.

315. Mountcastle, V. B., *Perceptual neuroscience: the cerebral cortex*. Cambridge, Mass.: Harvard University Press, 1998.

316. Laufs, H., K. Krakow, P. Sterzer, E. Eger, A. Beyerle, A. Salek-Haddadi, and A. Kleinschmidt, Electroencephalographic signatures of attentional and cognitive default modes in spontaneous brain activity fluctuations at rest. *Proc Natl Acad Sci U S A*, 2003. 100(19): pp. 11053–8.

317. Berridge, C. W., S. J. Bolen, M. S. Manley, and S. L. Foote, Modulation of forebrain electroencephalographic activity in halothane-anesthetized rat via actions of noradrenergic beta-receptors within the medial septal region. *J Neurosci*, 1996. 16(21): pp. 7010–20.

318. Duque, A., B. Balatoni, L. Detari, and L. Zaborszky, EEG correlation of the discharge properties of identified neurons in the basal forebrain. *J Neurophysiol*, 2000. 84(3): pp. 1627–35.

319. Gazzaniga, M. S., *The new cognitive neurosciences*. Cambridge, Mass.: The MIT Press, 2000.

320. Raichle, M., The neural correlates of consciousness: an analysis of cognitive skill learning, in *The cognitive neurosciences*, M. S. Gazzaniga and E. Bizzi, eds. Cambridge, Mass.: The MIT Press. pp. 1305–18.

321. Zeman, A., *Consciousness: a user's guide*. New Haven: Yale University Press, 2002.

322. Saint-Mleux, B., E. Eggermann, A. Bisetti, L. Bayer, D. Machard, B. E. Jones, M. Muhlethaler, and M. Serafin, Nicotinic enhancement of the noradrenergic inhibition of sleep-promoting neurons in the ventrolateral preoptic area. *J Neurosci*, 2004. 24(1): pp. 63–7.

323. McGinty, D., M. N. Alam, R. Szymusiak, M. Nakao, and M. Yamamoto, Hypothalamic sleep-promoting mechanisms: coupling to thermoregulation. *Arch Ital Biol*, 2001. 139(1–2): pp. 63–75.

324. Gallopin, T., P. Fort, E. Eggermann, B. Cauli, P. H. Luppi, J. Rossier, E. Audinat, M. Muhlethaler, and M. Serafin, Identification of sleep-promoting neurons in vitro. *Nature*, 2000. 404(6781): pp. 992–5.

325. Janig, W., The autonomic nervous system and its coordination by the brain, in *Handbook of affective sciences*, R. J. Davidson, K. R. Scherer, and H. H. Goldsmith, eds. Oxford: Oxford University Press, 2003.

326. Henry, J. P., Neuroendocrine patterns of emotional response, in *Biological foundations of emotion*, R. Plutchik and H. Kellerman, eds. Orlando: Academic Press, 1986.

327. Jordan, D., Autonomic changes in affective behavior, in *Central regulation of autonomic functions*, A. D. Loewy and K. M. Spyer, eds. New York: Oxford University Press, 1990.

328. Ogawa, S., E. Choleris, and D. W. Pfaff, Routes of genetic influences on aggressive behaviors in animals. *Ann N Y Acad Sci*, 2004, in press.

329. Ma, P. M., On the agonistic display of the Siamese fighting fish. II. The distribution, number and morphology of opercular display motoneurons. *Brain Behav Evol*, 1995. 45(6): pp. 314–26.

330. Gorlick, D. L., Neural pathway for aggressive display in Betta splendens: midbrain and hindbrain control of gill-cover erection behavior. *Brain Behav Evol*, 1990. 36(4): pp. 227–36.

331. Scott, J. P. and J. L. Fuller, *Genetics and the social behavior of the dog*. Chicago: University of Chicago Press, 1965.

332. Fuller, J. L., Genetics and emotions, in *Biological foundations of emotion*, R. Plutchik and H. Kellerman, eds. 1986, Orlando: Academic Press, 1986. pp. 199–217.

333. Cloninger, C. R., A systematic method for clinical description and classification of personality variants. *Arch Gen Psychiatry*, 1987. 44: pp. 573–88.

334. Cloninger, C. R., *Feeling good: the science of well-being*. Oxford: Oxford University Press, 2004.

335. Cloninger, C. R., D. M. Svrakic, and T. R. Przybeck, A psychobiological model of temperament and character. *Arch Gen Psychiatry*, 1993. 50(12): pp. 975–90.

336. Viding, E., Evidence for genetic influences on emotional behaviors in humans. *Ann N Y Acad Sci*, 2004, in press.

337. Kelly, J. P., Principles of the functional and anatomical organization of the nervous system, in *Principles of neural science*, E. R. Kandel and J. H. Schwartz, eds. New York: Elsevier, 1985. p. 979.

338. Powley, T. L., Central control of autonomic functions: the organization of the autonomic nervous system, in *Fundamental neuroscience*, M. J. Zigmond, F. E. Bloom, S. C. Landis, J. L. Roberts, and L. R. Squire, eds. Orlando: Academic Press, 1999.

339. McKenna, K. E., Some proposals regarding the organization of the central nervous system control of penile erection. *Neurosci Biobehav Rev*, 2000. 24(5): pp. 535–40.

340. Saper, C. B., The central autonomic nervous system: conscious visceral perception and autonomic pattern generation. *Annu Rev Neurosci*, 2002. 25: pp. 433–69.

341. Iverson, S., L. Iverson, and C. Saper, The autonomic nervous system and the hypo-

thalamus, in *Principles of neural science*, E. R. Kandel and J. H. Schwartz, eds. New York: McGraw-Hill, 2000.

342. Thompson, R. H., and L. W. Swanson, Structural characterization of a hypothalamic visceromotor pattern generator network. *Brain Res Rev*, 2003. 41(2–3): pp. 153–202.

343. Swanson, L. W., and P. E. Sawchenko, Hypothalamic integration: organization of the paraventricular and supraoptic nuclei. *Annu Rev Neurosci*, 1983. 6: pp. 269–324.

344. Conrad, L., and D. Pfaff, Efferents from medial basal forebrain and hypothalamus in the rat. II. An autoradiography study of the anterior hypothalamus. *J Comp Neurol*, 1976. 169: pp. 221–62.

345. Conrad, L., and D. Pfaff, Efferents from medial basal forebrain and hypothalamus in the rat. I. An autoradiography study of the medial preoptic area. *J Comp Neurol*, 1976. 169: pp. 185–220.

346. Saper, C. B., L. W. Swanson, and W. M. Cowan, The efferent connections of the ventromedial nucleus of the hypothalamus of the rat. *J Comp Neurol*, 1976. 169(4): pp. 409–42.

347. Saper, C. B., A. D. Loewy, L. W. Swanson, and W. M. Cowan, Direct hypothalamo-autonomic connections. *Brain Res*, 1976. 117(2): pp. 305–12.

348. Hosoya, Y., Y. Sugiura, N. Okado, A. D. Loewy, and K. Kohno, Descending input from the hypothalamic paraventricular nucleus to sympathetic preganglionic neurons in the rat. *Exp Brain Res*, 1991. 85(1): pp. 10–20.

349. Ross, A. R., and R. B. Malmo, Cardiovascular responses to rewarding brain stimulation. *Physiol Behav*, 1979. 22(5): pp. 1005–13.

350. Xu, Y., and T. L. Krukoff, Decrease in arterial pressure induced by adrenomedullin in the hypothalamic paraventricular nucleus is mediated by nitric oxide and GABA. *Regul Pept*, 2004. 119(1–2): pp. 21–30.

351. Kenney, M. J., M. L. Weiss, T. Mendes, Y. Wang, and R. J. Fels, Role of paraventricular nucleus in regulation of sympathetic nerve frequency components. *Am J Physiol Heart Circ Physiol*, 2003. 284(5): pp. H1710–20.

352. Kenney, M. J., M. L. Weiss, and J. R. Haywood, The paraventricular nucleus: an important component of the central neurocircuitry regulating sympathetic nerve outflow. *Acta Physiol Scand*, 2003. 177(1): pp. 7–15.

353. Harris, M. C., and A. D. Loewy, Neural regulation of vasopressin-containing hypothalamic neurons and the role of vasopressin in cardiovascular function, in *Central regulation of autonomic functions*, A. D. Loewy and K. M. Spyer, eds. New York: Oxford University Press, 1990.

354. Wang, J., M. Irnaten, P. Venkatesan, C. Evans, and D. Mendelowitz, Arginine vasopressin enhances GABAergic inhibition of cardiac parasympathetic neurons in the nucleus ambiguus. *Neuroscience*, 2002. 111(3): pp. 699–705.

355. Streefkerk, J. O., M. J. Mathy, M. Pfaffendorf, and P. A. van Zwieten, Vasopressin-induced presynaptic facilitation of sympathetic neurotransmission in the pithed rat. *J Hypertens*, 2002. 20(6): pp. 1175–80.

356. Yang, Z., and J. H. Coote, The influence of vasopressin on tonic activity of cardiovascular neurones in the ventrolateral medulla of the hypertensive rat. *Auton Neurosci*, 2003. 104(2): pp. 83–7.

357. Sansone, G. R., and B. R. Komisaruk, Evidence that oxytocin is an endogenous stimulator of autonomic sympathetic preganglionics: the pupillary dilatation response to vaginocervical stimulation in the rat. *Brain Res*, 2001. 898(2): pp. 265–71.

358. Haxhiu, M. A., F. Tolentino-Silva, G. Pete, P. Kc, and S. O. Mack, Monoaminergic neurons, chemosensation and arousal. *Respir Physiol*, 2001. 129(1–2): pp. 191–209.

359. Mack, S. O., P. Kc, M. Wu, B. R. Coleman, F. P. Tolentino-Silva, and M. A. Haxhiu, Paraventricular oxytocin neurons are involved in neural modulation of breathing. *J Appl Physiol*, 2002. 92(2): pp. 826–34.

360. Loewy, A. D., Central autonomic pathways, in *Central regulation of autonomic functions*, A. D. Loewy and K. M. Spyer, eds. New York: Oxford University Press, 1990.

361. Spyer, K. M., The central nervous organization of reflex circulatory control, in *Central regulation of autonomic functions*, A. D. Loewy and K. M. Spyer, eds. New York: Oxford University Press, 1990.

362. Guyenet, P. G., A. M. Schreihofer, and R. L. Stornetta, Regulation of sympathetic tone and arterial pressure by the rostral ventrolateral medulla after depletion of C1 cells in rats. *Ann N Y Acad Sci*, 2001. 940: pp. 259–69.

363. Kerman, I. A., L. W. Enquist, S. J. Watson, and B. J. Yates, Brainstem substrates of sympatho-motor circuitry identified using trans-synaptic tracing with pseudorabies virus recombinants. *J Neurosci*, 2003. 23(11): pp. 4657–66.

364. Yasuda, T., T. Masaki, T. Sakata, and H. Yoshimatsu, Hypothalamic neuronal histamine regulates sympathetic nerve activity and expression of uncoupling protein 1 mRNA in brown adipose tissue in rats. *Neuroscience*, 2004. 125(3): pp. 535–40.

365. Hugelin, A., Forebrain and midbrain influence on respiration, in *Handbook of physiology: respiration*, E. A. Fishman, ed. Washington, D.C.: American Physiological Society, 1964. pp. 69–87.

366. Aggleton, J. P., and M. Mishkin, The amygdala: sensory gateway to the emotions, in *Biological foundations of emotion*, R. Plutchik and H. Kellerman, eds. Orlando: Academic Press, 1986.

367. LeDoux, J. E., Emotion circuits in the brain. *Annu Rev Neurosci*, 2000. 23: pp. 155–84.

368. Benarroch, E. E., Functional anatomy of the central autonomic nervous system, in *Handbook of clinical neurology*, O. Appenzeller, ed. Philadelphia: Saunders, 1999.

369. Taylor, E. W., D. Jordan, and J. H. Coote, Central control of the cardiovascular and respiratory systems and their interactions in vertebrates. *Physiol Rev*, 1999. 79(3): pp. 855–916.

370. Feldman, J. L., G. S. Mitchell, and E. E. Nattie, Breathing: rhythmicity, plasticity, chemosensitivity. *Annu Rev Neurosci*, 2003. 26: pp. 239–66.

371. Mead, J., and E. Agostoni, Dynamics of breathing, in *Handbook of physiology: respiration*, W. O. Fenn and H. Rahn, eds. Washington, D. C.: American Physiological Society, 1964. Chapter 14.

372. Wang, S. C., and S. H. Ngai, General organization of central respiratory mechanisms, in *Handbook of physiology: respiration*, W. O. Fenn and H. Rahn, eds. Washington, D. C.: American Physiological Society, 1964. Chapter 19.

373. Dempsey, J. A., E. B. Olson, and J. Skatrud, Hormones and neurochemicals in the regulation of breathing, in *Handbook of physiology: respiration*, E. A. Fishman et al., eds. Washington, D.C.: American Physiological Society, 1964.

374. Krout, K. E., T. C. Mettenleiter, and A. D. Loewy, Single CNS neurons link both central motor and cardiosympathetic systems: a double-virus tracing study. *Neuroscience*, 2003. 118(3): pp. 853–66.

375. Loewy, A. D., and M. A. Haxhiu, CNS cell groups projecting to pancreatic parasympathetic preganglionic neurons. *Brain Res*, 1993. 620(2): pp. 323–30.

376. Westerhaus, M. J., and A. D. Loewy, Central representation of the sympathetic nervous system in the cerebral cortex. *Brain Res*, 2001. 903(1–2): pp. 117–27.

377. Meyer, G., B. P. Hauffa, M. Schedlowski, C. Pawlak, M. A. Stadler, and M. S. Exton, Casino gambling increases heart rate and salivary cortisol in regular gamblers. *Biol Psychiatry*, 2000. 48(9): pp. 948–53.

378. Kahneman, D., *Attention and effort*. Prentice-Hall Series in Experimental Psychology. Englewood Cliffs, N.J.: Prentice-Hall, 1973.

379. Obrist, P. A., R. A. Webb, J. R. Sutterer, and J. L. Howard, The cardiac-somatic relationship: some reformulations. *Psychophysiology*, 1970. 6(5): pp. 569–87.

380. Chase, W. G., F. K. Graham, and D. T. Graham, Components of HR response in anticipation of reaction time and exercise tasks. *J Exp Psychol*, 1968. 76(4): pp. 642–8.

381. Obrist, P. A., R. A. Webb, and J. R. Sutterer, Heart rate and somatic changes during aversive conditioning and a simple reaction time task. *Psychophysiology*, 1969. 5(6): pp. 696–723.

382. Obrist, P. A., Heart rate and somatic-motor coupling during classical aversive conditioning in humans. *J Exp Psychol*, 1968. 77(2): pp. 180–93.

383. Wilson, R. S., Autonomic changes produced by noxious and innocuous stimulation. *J Comp Physiol Psychol*, 1964. 58: pp. 290–5.

384. Deane, G. E., Human heart rate responses during experimentally induced anxiety. *J Exp Psychol*, 1961. 61: pp. 489–93.

385. Cohen, M. J., and H. J. Johnson, Relationship between heart rate and muscular activity within a classical conditioning paradigm. *J Exp Psychol*, 1971. 90(2): pp. 222–6.

386. Connor, W. H., and P. J. Lang, Cortical slow-wave and cardiac rate responses in stimulus orientation and reaction time conditions. *J Exp Psychol*, 1969. 82(2): pp. 310–20.

387. Obrist, P. A., K. C. Light, J. A. McCubbin, J. S. Hutcheson, and J. L. Hoffer, Pulse transit time: relationship to blood pressure and myocardial performance. *Psychophysiology*, 1979. 16(3): pp. 292–301.

388. Webb, R. A., and P. A. Obrist, The physiological concomitants of reaction time performance as a function of preparatory interval and preparatory interval series. *Psychophysiology*, 1970. 6(4): pp. 389–403.

389. Morin, D., and D. Viala, Coordinations of locomotor and respiratory rhythms in vitro are critically dependent on hindlimb sensory inputs. *J Neurosci*, 2002. 22(11): pp. 4756–65.

390. Potts, J. T., Neural circuits controlling cardiorespiratory responses: baroreceptor and somatic afferents in the nucleus tractus solitarius. *Clin Exp Pharmacol Physiol*, 2002. 29(1–2): pp. 103–11.

391. Lacey, J. I., Somatic response patterning and stress: some revisions of activation theory, in *Psychological stress, issues in research*, M. H. Appley, R. Trumbull, eds. New York: Appleton-Century-Crofts, 1967.

392. Libby, W. L., Jr., B. C. Lacey, and J. I. Lacey, Pupillary and cardiac activity during visual attention. *Psychophysiology*, 1973. 10(3): pp. 270–94.

393. Tursky, B., G. E. Schwartz, and A. Crider, Differential patterns of heart rate and skin resistance during a digit-transformation task. *J Exp Psychol*, 1970. 83(3): pp. 451–7.

394. Schommer, N. C., D. H. Hellhammer, and C. Kirschbaum, Dissociation between reactivity of the hypothalamus-pituitary-adrenal axis and the sympathetic-adrenal-medullary system to repeated psychosocial stress. *Psychosom Med*, 2003. 65(3): pp. 450–60.

395. Hunt, P. S., and J. S. Phillips, Postnatal binge ethanol exposure affects habituation of the cardiac orienting response to an olfactory stimulus in preweanling rats. *Alcohol Clin Exp Res*, 2004. 28(1): pp. 123–30.

396. Mauss, I. B., F. H. Wilhelm, and J. J. Gross, Autonomic recovery and habituation in social anxiety. *Psychophysiology*, 2003. 40(4): pp. 648–53.

397. Papka, R., B. Srinivasan, K. Miller, and S. Hayashi, Localization of estrogen receptor protein and estrogen receptor messenger RNA in peripheral autonomic and sensory neurons. *Neuroscience*, 1997. 79: pp. 1153–63.

398. Zoubina, E. V., and P. G. Smith, Distributions of estrogen receptors alpha and beta in sympathetic neurons of female rats: enriched expression by uterine innervation. *J Neurobiol*, 2002. 52(1): pp. 14–23.

399. Zoubina, E. V., and P. G. Smith, Expression of estrogen receptors alpha and beta by sympathetic ganglion neurons projecting to the proximal urethra of female rats. *J Urol*, 2003. 169(1): pp. 382–5.

400. Mendelsohn, M., and R. Karas, The protective effects of estrogen on the cardiovascular system. *N Engl J Med*, 1999. 340: pp. 1801–11.

401. Jankowski, M., G. Rachelska, W. Donghao, S. M. McCann, and J. Gutkowska, Estrogen receptors activate atrial natriuretic peptide in the rat heart. *Proc Natl Acad Sci U S A*, 2001. 98(20): pp. 11765–70.

402. Nickenig, G., K. Strehlow, S. Wassmann, A. T. Baumer, K. Albory, H. Sauer, and M. Bohm, Differential effects of estrogen and progesterone on AT(1) receptor gene expression in vascular smooth muscle cells. *Circulation*, 2000. 102(15): pp. 1828–33.

403. Ciriello, J., C. V. R. de Oliveira, M. Masoumeh, M. Ganjkhani, and Z. Li, Estrogen alters the cardiovascular responses to activation of rostral ventrolateral medulla in the female. Society for Neuroscience Abstracts, 2002: p. 768.6.

404. Rhodes, C., J. I. Morrell, and D. Pfaff, Estrogen- neurophysin-containing hypothalamic magnocellular neurons in the vasopressin-deficient (Brattleboro) rat: a study combining steroid autoradiography and immunocytochemistry. *J Neurosci*, 1982. 2: pp. 1718–24.

405. Veronneau-Longueville, F., O. Rampin, M. J. Freund-Mercier, Y. Tang, A. Calas, L. Marson, K. E. McKenna, M. E. Stoeckel, G. Benoit, and F. Giuliano, Oxytocinergic innervation of autonomic nuclei controlling penile erection in the rat. *Neuroscience*, 1999. 93(4): pp. 1437–47.

406. Meston, C. M., I. V. Moe, and B. B. Gorzalka, Effects of sympathetic inhibition on receptive, proceptive, and rejection behaviors in the female rat. *Physiol Behav*, 1996. 59(3): pp. 537–42.

407. Hague, C., Z. Chen, M. Uberti, and K. P. Minneman, Alpha(1)-adrenergic receptor subtypes: non-identical triplets with different dancing partners? *Life Sci*, 2003. 74(4): pp. 411–8.

408. Piascik, M. T., and D. M. Perez, Alpha(1)-adrenergic receptors: new insights and directions. *J Pharmacol Exp Ther*, 2001. 298(2): pp. 403–10.

409. Garland, E. M., and I. Biaggioni, Genetic polymorphisms of adrenergic receptors. *Clin Auton Res*, 2001. 11(2): pp. 67–78.

410. Koch, W. J., R. J. Lefkowitz, and H. A. Rockman, Functional consequences of altering myocardial adrenergic receptor signaling. *Annu Rev Physiol*, 2000. 62: pp. 237–60.

411. Etgen, A. M., and J. C. Morales, Somatosensory stimuli evoke norepinephrine release in the anterior ventromedial hypothalamus of sexually receptive female rats. *J Neuroendocrinol*, 2002. 14(3): pp. 213–8.

412. Sabban, E. L., and R. Kvetnansky, Stress-triggered activation of gene expression in catecholaminergic systems: dynamics of transcriptional events. *Trends Neurosci*, 2001. 24(2): pp. 91–8.

413. Serova, L., M. Rivkin, A. Nakashima, and E. L. Sabban, Estradiol stimulates gene expression of norepinephrine biosynthetic enzymes in rat locus coeruleus. *Neuroendocrinology*, 2002. 75(3): pp. 193–200.

414. Small, K. M., D. W. McGraw, and S. B. Liggett, Pharmacology and physiology of human adrenergic receptor polymorphisms. *Annu Rev Pharmacol Toxicol*, 2003. 43: pp. 381–411.

415. Graves, L. A., K. Hellman, S. Veasey, J. A. Blendy, A. I. Pack, and T. Abel, Genetic evi-

dence for a role of CREB in sustained cortical arousal. *J Neurophysiol*, 2003. 90(2): pp. 1152–9.

416. Uhl, G. R., F. S. Hall, and I. Sora, Cocaine, reward, movement and monoamine transporters. *Mol Psychiatry*, 2002. 7(1): pp. 21–6.

417. Gainetdinov, R. R., T. D. Sotnikova, and M. G. Caron, Monoamine transporter pharmacology and mutant mice. *Trends Pharmacol Sci*, 2002. 23(8): pp. 367–73.

418. Garland, E. M., M. K. Hahn, T. P. Ketch, N. R. Keller, C. H. Kim, K. S. Kim, I. Biaggioni, J. R. Shannon, R. D. Blakely, and D. Robertson, Genetic basis of clinical catecholamine disorders. *Ann N Y Acad Sci*, 2002. 971: pp. 506–14.

419. Kalani, M. Y., N. Vaidehi, S. E. Hall, R. J. Trabanino, P. L. Freddolino, M. A. Kalani, W. B. Floriano, V. W. Kam, and W. A. Goddard, III, The predicted 3D structure of the human D2 dopamine receptor and the binding site and binding affinities for agonists and antagonists. *Proc Natl Acad Sci U S A*, 2004. 101(11): pp. 3815–20.

420. Noble, E. P., D2 dopamine receptor gene in psychiatric and neurologic disorders and its phenotypes. *Am J Med Genet*, 2003. 116B(1): pp. 103–25.

421. Le Crom, S., M. Kapsimali, P. O. Barome, and P. Vernier, Dopamine receptors for every species: gene duplications and functional diversification in Craniates. *J Struct Funct Genomics*, 2003. 3(1–4): pp. 161–76.

422. Glatt, C. E., and V. I. Reus, Pharmacogenetics of monoamine transporters. *Pharmacogenomics*, 2003. 4(5): pp. 583–96.

423. Ueno, S., Genetic polymorphisms of serotonin and dopamine transporters in mental disorders. *J Med Invest*, 2003. 50(1–2): pp. 25–31.

424. Krause, K. H., S. H. Dresel, J. Krause, C. la Fougere, and M. Ackenheil, The dopamine transporter and neuroimaging in attention deficit hyperactivity disorder. *Neurosci Biobehav Rev*, 2003. 27(7): pp. 605–13.

425. Tan, S., B. Hermann, and E. Borrelli, Dopaminergic mouse mutants: investigating the roles of the different dopamine receptor subtypes and the dopamine transporter. *Int Rev Neurobiol*, 2003. 54: pp. 145–97.

426. Lesch, K. P., D. Bengel, A. Heils, S. Z. Sabol, B. D. Greenberg, S. Petri, J. Benjamin, C. R. Muller, D. H. Hamer, and D. L. Murphy, Association of anxiety-related traits with a polymorphism in the serotonin transporter gene regulatory region. *Science*, 1996. 274(5292): pp. 1527–31.

427. Hariri, A. R., V. S. Mattay, A. Tessitore, B. Kolachana, F. Fera, D. Goldman, M. F. Egan, and D. R. Weinberger, Serotonin transporter genetic variation and the response of the human amygdala. *Science*, 2002. 297(5580): pp. 400–3.

428. Caspi, A., K. Sugden, T. E. Moffitt, A. Taylor, I. W. Craig, H. Harrington, J. McClay, J. Mill, J. Martin, A. Braithwaite, and R. Poulton, Influence of life stress on depression: moderation by a polymorphism in the 5-HTT gene. *Science*, 2003. 301(5631): pp. 386–9.

429. Sanders, A. R., J. Duan, and P. V. Gejman, DNA variation and psychopharmacology of the human serotonin receptor 1B (HTR1B) gene. *Pharmacogenomics*, 2002. 3(6): pp. 745–62.

430. Wada, H., N. Inagaki, A. Yamatodani, and T. Watanabe, Is the histaminergic neuron system a regulatory center for whole-brain activity? *Trends Neurosci*, 1991. 14(9): pp. 415–8.

431. Brown, R. E., D. R. Stevens, and H. L. Haas, The physiology of brain histamine. *Prog Neurobiol*, 2001. 63(6): pp. 637–72.

432. Lin, J. S., K. Sakai, and M. Jouvet, Hypothalamo-preoptic histaminergic projections in sleep-wake control in the cat. *Eur J Neurosci*, 1994. 6(4): pp. 618–25.

433. Lin, J. S., Y. Hou, K. Sakai, and M. Jouvet, Histaminergic descending inputs to the

mesopontine tegmentum and their role in the control of cortical activation and wakefulness in the cat. *J Neurosci*, 1996. 16(4): pp. 1523–37.

434. Scammell, T. E., I. V. Estabrooke, M. T. McCarthy, R. M. Chemelli, M. Yanagisawa, M. S. Miller, and C. B. Saper, Hypothalamic arousal regions are activated during modafinil-induced wakefulness. *J Neurosci*, 2000. 20(22): pp. 8620–8.

435. Parmentier, R., H. Ohtsu, Z. Djebbara-Hannas, J. L. Valatx, T. Watanabe, and J. S. Lin, Anatomical, physiological, and pharmacological characteristics of histidine decarboxylase knock-out mice: evidence for the role of brain histamine in behavioral and sleep-wake control. *J Neurosci*, 2002. 22(17): pp. 7695–711.

436. Yanai, K., L. Z. Son, M. Endou, E. Sakurai, O. Nakagawasai, T. Tadano, K. Kisara, I. Inoue, and T. Watanabe, Behavioral characterization and amounts of brain monoamines and their metabolites in mice lacking histamine H1 receptors. *Neuroscience*, 1998. 87(2): pp. 479–87.

437. Lin, J. S., K. Kitahama, P. Fort, P. Panula, R. M. Denney, and M. Jouvet, Histaminergic system in the cat hypothalamus with reference to type B monoamine oxidase. *J Comp Neurol*, 1993. 330(3): pp. 405–20.

438. Yamanaka, A., N. Tsujino, H. Funahashi, K. Honda, J. L. Guan, Q. P. Wang, M. Tominaga, K. Goto, S. Shioda, and T. Sakurai, Orexins activate histaminergic neurons via the orexin 2 receptor. *Biochem Biophys Res Commun*, 2002. 290(4): pp. 1237–45.

439. Eriksson, K. S., O. Sergeeva, R. E. Brown, and H. L. Haas, Orexin/hypocretin excites the histaminergic neurons of the tuberomammillary nucleus. *J Neurosci*, 2001. 21(23): pp. 9273–9.

440. Lipsky, R. H., and D. Goldman, Genomics and variation of ionotropic glutamate receptors. *Ann N Y Acad Sci*, 2003. 1003: pp. 22–35.

441. Madden, D. R., The structure and function of glutamate receptor ion channels. *Nat Rev Neurosci*, 2002. 3(2): pp. 91–101.

442. Liu, L., T. P. Wong, M. F. Pozza, K. Lingenhoehl, Y. Wang, M. Sheng, Y. P. Auberson, and Y. T. Wang, Role of NMDA receptor subtypes in governing the direction of hippocampal synaptic plasticity. *Science*, 2004. 304(5673): pp. 1021–4.

443. Szekely, J. I., K. Torok, and G. Mate, The role of ionotropic glutamate receptors in nociception with special regard to the AMPA binding sites. *Curr Pharm Des*, 2002. 8(10): pp. 887–912.

444. Coyle, J. T., G. Tsai, and D. C. Goff, Ionotropic glutamate receptors as therapeutic targets in schizophrenia. *Curr Drug Targets CNS Neurol Disord*, 2002. 1(2): pp. 183–9.

445. Pin, J. P., and F. Acher, The metabotropic glutamate receptors: structure, activation mechanism and pharmacology. *Curr Drug Targets CNS Neurol Disord*, 2002. 1(3): pp. 297–317.

446. Conn, P. J., Physiological roles and therapeutic potential of metabotropic glutamate receptors. *Ann N Y Acad Sci*, 2003. 1003: pp. 12–21.

447. Coutinho, V., and T. Knopfel, Metabotropic glutamate receptors: electrical and chemical signaling properties. *Neuroscientist*, 2002. 8(6): pp. 551–61.

448. Raiteri, L., M. Raiteri, and G. Bonanno, Coexistence and function of different neurotransmitter transporters in the plasma membrane of CNS neurons. *Prog Neurobiol*, 2002. 68(4): pp. 287–309.

449. Slotboom, D. J., W. N. Konings, and J. S. Lolkema, Structural features of the glutamate transporter family. *Microbiol Mol Biol Rev*, 1999. 63(2): pp. 293–307.

450. Danbolt, N. C., Glutamate uptake. *Prog Neurobiol*, 2001. 65(1): pp. 1–105.

451. O'Shea, R. D., Roles and regulation of glutamate transporters in the central nervous system. *Clin Exp Pharmacol Physiol*, 2002. 29(11): pp. 1018–23.

452. Dunwiddie, T. V., and S. A. Masino, The role and regulation of adenosine in the central nervous system. *Annu Rev Neurosci*, 2001. 24: pp. 31–55.

453. Fredholm, B. B., K. A. Jacobson, K. N. Klotz, and J. Linden, International Union of Pharmacology. XXV. Nomenclature and classification of adenosine receptors. *Pharmacol Rev*, 2001. 53(4): pp. 527–52.

454. Song, W. J., T. Thatch, and D. J. Surmeier, Adenosine receptor expression and modulation of Ca(2+) channels in rat striatal cholinergic interneurons. *J Neurophysiol*, 2000. 83(1): pp. 322–32.

455. Nishi, A., F. Liu, S. Matsuyama, M. Hamada, H. Higashi, A. C. Nairn, and P. Greengard, Metabotropic mGlu5 receptors regulate adenosine A2A receptor signaling. *Proc Natl Acad Sci U S A*, 2003. 100(3): pp. 1322–7.

456. Scammell, T. E., D. Y. Gerashchenko, T. Mochizuki, M. T. McCarthy, I. V. Estabrooke, C. A. Sears, C. B. Saper, Y. Urade, and O. Hayaishi, An adenosine A2a agonist increases sleep and induces Fos in ventrolateral preoptic neurons. *Neuroscience*, 2001. 107(4): pp. 653–63.

457. Luscher, B., and J. M. Fritschy, Subcellular localization and regulation of GABA-A receptors and associated proteins. *Int Rev Neurobiol*, 2001. 48: pp. 31–64.

458. Porjesz, B., L. Almasy, H. J. Edenberg, K. Wang, D. B. Chorlian, T. Foroud, A. Goate, J. P. Rice, S. J. O'Connor, J. Rohrbaugh, S. Kuperman, L. O. Bauer, R. R. Crowe, M. A. Schuckit, V. Hesselbrock, P. M. Conneally, J. A. Tischfield, T. K. Li, T. Reich, and H. Begleiter, Linkage disequilibrium between the beta frequency of the human EEG and a GABA-A receptor gene locus. *Proc Natl Acad Sci U S A*, 2002. 99(6): pp. 3729–33.

459. Krasowski, M. D., R. A. Harris, and N. L. Harrison, Allosteric modulation of GABAa receptor function by general anesthetics and alcohols. Section: Pharmacology of GABA and glycine neurotransmission. In: Handbook of experimental pharmacology; v. 150, ed. B. E. Alger and H. Mèohler. Berlin: Springer-Verlag, 2001. Chapter 5.

460. Nishikawa, K., A. Jenkins, I. Paraskevakis, and N. L. Harrison, Volatile anesthetic actions on the GABAA receptors: contrasting effects of alpha 1(S270) and beta 2(N265) point mutations. *Neuropharmacology*, 2002. 42(3): pp. 337–45.

461. Kopp, C., U. Rudolph, K. Low, and I. Tobler, Modulation of rhythmic brain activity by diazepam: GABA(A) receptor subtype and state specificity. *Proc Natl Acad Sci U S A*, 2004. 101(10): pp. 3674–9.

462. Steffensen, S. C., S. H. Stobbs, D. W. Allison, M. B. Lassen, J. E. Brown, and S. J. Henriksen, Ventral tegmental area GABA neurons form a network of dopamine-sensitive electrical synapses: role in brain stimulation reward. Society for Neuroscience Abstracts, 2003: p. 679.4.

463. Willie, J. T., R. M. Chemelli, C. M. Sinton, and M. Yanagisawa, To eat or to sleep? Orexin in the regulation of feeding and wakefulness. *Annu Rev Neurosci*, 2001. 24: pp. 429–58.

464. Sutcliffe, J. G., and L. de Lecea, The hypocretins: setting the arousal threshold. *Nat Rev Neurosci*, 2002. 3(5): pp. 339–49.

465. Taheri, S., J. M. Zeitzer, and E. Mignot, The role of hypocretins (orexins) in sleep regulation and narcolepsy. *Annu Rev Neurosci*, 2002. 25: pp. 283–313.

466. Mieda, M., J. T. Willie, J. Hara, C. M. Sinton, T. Sakurai, and M. Yanagisawa, Orexin peptides prevent cataplexy and improve wakefulness in an orexin neuron-ablated model of narcolepsy in mice. *Proc Natl Acad Sci U S A*, 2004. 101(13): pp. 4649–54.

467. Nambu, T., T. Sakurai, K. Mizukami, Y. Hosoya, M. Yanagisawa, and K. Goto, Distribution of orexin neurons in the adult rat brain. *Brain Res*, 1999. 827(1–2): pp. 243–60.

468. Hervieu, G. J., J. E. Cluderay, D. C. Harrison, J. C. Roberts, and R. A. Leslie, Gene ex-

pression and protein distribution of the orexin-1 receptor in the rat brain and spinal cord. *Neuroscience*, 2001. 103(3): pp. 777–97.

469. Sakurai, T., A. Amemiya, M. Ishii, I. Matsuzaki, R. M. Chemelli, H. Tanaka, S. C. Williams, J. A. Richardson, G. P. Kozlowski, S. Wilson, J. R. Arch, R. E. Buckingham, A. C. Haynes, S. A. Carr, R. S. Annan, D. E. McNulty, W. S. Liu, J. A. Terrett, N. A. Elshourbagy, D. J. Bergsma, and M. Yanagisawa, Orexins and orexin receptors: a family of hypothalamic neuropeptides and G protein-coupled receptors that regulate feeding behavior. *Cell*, 1998. 92(4): pp. 573–85.

470. Diano, S., B. Horvath, H. F. Urbanski, P. Sotonyi, and T. L. Horvath, Fasting activates the nonhuman primate hypocretin (orexin) system and its postsynaptic targets. *Endocrinology*, 2003. 144(9): pp. 3774–8.

471. Kotz, C. M., J. A. Teske, J. A. Levine, and C. Wang, Feeding and activity induced by orexin A in the lateral hypothalamus in rats. *Regul Pept*, 2002. 104(1–3): pp. 27–32.

472. Yamanaka, A., C. T. Beuckmann, J. T. Willie, J. Hara, N. Tsujino, M. Mieda, M. Tominaga, K. Yagami, F. Sugiyama, K. Goto, M. Yanagisawa, and T. Sakurai, Hypothalamic orexin neurons regulate arousal according to energy balance in mice. *Neuron*, 2003. 38(5): pp. 701–13.

473. Peyron, C., D. K. Tighe, A. N. van den Pol, L. de Lecea, H. C. Heller, J. G. Sutcliffe, and T. S. Kilduff, Neurons containing hypocretin (orexin) project to multiple neuronal systems. *J Neurosci*, 1998. 18(23): pp. 9996–10015.

474. Eriksson, K. S., O. A. Sergeeva, O. Selbach, and H. L. Haas, Orexin (hypocretin)/dynorphin neurons control GABAergic inputs to tuberomammillary neurons. *Eur J Neurosci*, 2004. 19(5): pp. 1278–84.

475. Sakamoto, F., S. Yamada, and Y. Ueta, Centrally administered orexin-A activates corticotropin-releasing factor-containing neurons in the hypothalamic paraventricular nucleus and central amygdaloid nucleus of rats: possible involvement of central orexins on stress-activated central CRF neurons. *Regul Pept*, 2004. 118(3): pp. 183–91.

476. Sato-Suzuki, I., I. Kita, Y. Seki, M. Oguri, and H. Arita, Cortical arousal induced by microinjection of orexins into the paraventricular nucleus of the rat. *Behav Brain Res*, 2002. 128(2): pp. 169–77.

477. Eggermann, E., M. Serafin, L. Bayer, D. Machard, B. Saint-Mleux, B. E. Jones, and M. Muhlethaler, Orexins/hypocretins excite basal forebrain cholinergic neurones. *Neuroscience*, 2001. 108(2): pp. 177–81.

478. Hagan, J. J., R. A. Leslie, S. Patel, M. L. Evans, T. A. Wattam, S. Holmes, C. D. Benham, S. G. Taylor, C. Routledge, P. Hemmati, R. P. Munton, T. E. Ashmeade, A. S. Shah, J. P. Hatcher, P. D. Hatcher, D. N. Jones, M. I. Smith, D. C. Piper, A. J. Hunter, R. A. Porter, and N. Upton, Orexin A activates locus coeruleus cell firing and increases arousal in the rat. *Proc Natl Acad Sci U S A*, 1999. 96(19): pp. 10911–6.

479. van den Pol, A. N., P. K. Ghosh, R. J. Liu, Y. Li, G. K. Aghajanian, and X. B. Gao, Hypocretin (orexin) enhances neuron activity and cell synchrony in developing mouse GFP-expressing locus coeruleus. *J Physiol*, 2002. 541(Pt 1): pp. 169–85.

480. Russell, S. H., C. J. Small, C. L. Dakin, C. R. Abbott, D. G. Morgan, M. A. Ghatei, and S. R. Bloom, The central effects of orexin-A in the hypothalamic-pituitary-adrenal axis in vivo and in vitro in male rats. *J Neuroendocrinol*, 2001. 13(6): pp. 561–6.

481. Date, Y., Y. Ueta, H. Yamashita, H. Yamaguchi, S. Matsukura, K. Kangawa, T. Sakurai, M. Yanagisawa, and M. Nakazato, Orexins, orexigenic hypothalamic peptides, interact with autonomic, neuroendocrine and neuroregulatory systems. *Proc Natl Acad Sci U S A*, 1999. 96(2): pp. 748–53.

482. Yang, B., W. K. Samson, and A. V. Ferguson, Excitatory effects of orexin-A on nucleus

tractus solitarius neurons are mediated by phospholipase C and protein kinase C. *J Neurosci*, 2003. 23(15): pp. 6215–22.

483. Mong, J. A., N. Devidze, D. E. Frail, L. T. O'Connor, M. Samuel, E. Choleris, S. Ogawa, and D. W. Pfaff, Estradiol differentially regulates lipocalin-type prostaglandin D synthase transcript levels in the rodent brain: Evidence from high-density oligonucleotide arrays and in situ hybridization. *Proc Natl Acad Sci U S A*, 2003. 100(1): pp. 318–23.

484. Matsumura, H., T. Nakajima, T. Osaka, S. Satoh, K. Kawase, E. Kubo, S. S. Kantha, K. Kasahara, and O. Hayaishi, Prostaglandin D2-sensitive, sleep-promoting zone defined in the ventral surface of the rostral basal forebrain. *Proc Natl Acad Sci U S A*, 1994. 91(25): pp. 11998–2002.

485. Mizoguchi, A., N. Eguchi, K. Kimura, Y. Kiyohara, W. M. Qu, Z. L. Huang, T. Mochizuki, M. Lazarus, T. Kobayashi, T. Kaneko, S. Narumiya, Y. Urade, and O. Hayaishi, Dominant localization of prostaglandin D receptors on arachnoid trabecular cells in mouse basal forebrain and their involvement in the regulation of non-rapid eye movement sleep. *Proc Natl Acad Sci U S A*, 2001. 98(20): pp. 11674–9.

486. Mong, J. A., N. Devidze, A. Goodwillie, and D. W. Pfaff, Reduction of lipocalin-type prostaglandin D synthase in the preoptic area of female mice mimics estradiol effects on arousal and sex behavior. *Proc Natl Acad Sci U S A*, 2003. 100(25): pp. 15206–11.

487. Gerashchenko, D., Y. Okano, Y. Urade, S. Inoue, and O. Hayaishi, Strong rebound of wakefulness follows prostaglandin D2- or adenosine A2a receptor agonist-induced sleep. *J Sleep Res*, 2000. 9(1): pp. 81–7.

488. Hayaishi, O., Molecular genetic studies on sleep-wake regulation, with special emphasis on the prostaglandin D(2) system. *J Appl Physiol*, 2002. 92(2): pp. 863–8.

489. Huang, Z. L., Y. Sato, T. Mochizuki, T. Okada, W. M. Qu, A. Yamatodani, Y. Urade, and O. Hayaishi, Prostaglandin E2 activates the histaminergic system via the EP4 receptor to induce wakefulness in rats. *J Neurosci*, 2003. 23(14): pp. 5975–83.

490. Low-Zeddies, S. S., and J. S. Takahashi, Chimera analysis of the Clock mutation in mice shows that complex cellular integration determines circadian behavior. *Cell*, 2001. 105(1): pp. 25–42.

491. Kopp, C., U. Albrecht, B. Zheng, and I. Tobler, Homeostatic sleep regulation is preserved in mPer1 and mPer2 mutant mice. *Eur J Neurosci*, 2002. 16(6): pp. 1099–106.

492. Valentino, R. J., and E. van Bockstaele, Corticotropin-releasing factor: putative neurotransmitter actions of a neurohormone, in *Hormones, brain, and behavior*, D. W. Pfaff, A. Arnold, A.Etgen, S. Fahrbach, and R. Rubin, eds. San Diego: Academic Press/Elsevier, 2002. pp. 81–105.

493. Muller, M. B., M. E. Keck, T. Steckler, and F. Holsboer, Genetics of endocrine-behavior interactions, in *Hormones, brain, and behavior*, D. W. Pfaff, A. Arnold, A.Etgen, S. Fahrbach, and R. Rubin, eds. San Diego: Academic Press/Elsevier, 2002. pp. 263–303.

494. Chang, F. C., and M. R. Opp, Pituitary CRH receptor blockade reduces waking in the rat. *Physiol Behav*, 1999. 67(5): pp. 691–6.

495. Heinrichs, S. C., and M. Joppa, Dissociation of arousal-like from anxiogenic-like actions of brain corticotropin-releasing factor receptor ligands in rats. *Behav Brain Res*, 2001. 122(1): pp. 43–50.

496. Chang, F. C., and M. R. Opp, A corticotropin-releasing hormone antisense oligodeoxynucleotide reduces spontaneous waking in the rat. *Regul Pept*, 2004. 117(1): pp. 43–52.

497. Weninger, S. C., A. J. Dunn, L. J. Muglia, P. Dikkes, K. A. Miczek, A. H. Swiergiel, C. W. Berridge, and J. A. Majzoub, Stress-induced behaviors require the corticotropin-

releasing hormone (CRH) receptor, but not CRH. *Proc Natl Acad Sci U S A*, 1999. 96(14): pp. 8283–8.

498. Hauptman, J. G., G. K. DeJong, K. A. Blasko, and I. H. Chaudry, Measurement of hepatocellular function, cardiac output, effective blood volume, and oxygen saturation in rats. *Am J Physiol*, 1989. 257(2 Pt 2): pp. R439–44.

499. Richard, D., and C. W. Bourque, Synaptic control of rat supraoptic neurones during osmotic stimulation of the organum vasculosum lamina terminalis in vitro. *J Physiol*, 1995. 489 (Pt 2): pp. 567–77.

500. Ozaki, Y., M. Nomura, J. Saito, C. E. Luedke, L. J. Muglia, T. Matsumoto, S. Ogawa, Y. Ueta, and D. W. Pfaff, Expression of the arginine vasopressin gene in response to salt loading in oxytocin gene knockout mice. *J Neuroendocrinol*, 2004. 16(1): pp. 39–44.

501. Bauer, M., and C. Whybrow, Thyroid hormone, brain and behavior, in *Hormones, brain, and behavior*, D. W. Pfaff, A. Arnold, A.Etgen, S. Fahrbach, and R. Rubin, eds. San Diego: Academic Press/Elsevier, 2002. pp. 239–65.

502. Joffe, R. T., Hypothalamic-pituitary-thyroid axis, in *Hormones, brain, and behavior*, D. W. Pfaff, A. Arnold, A.Etgen, S. Fahrbach, and R. Rubin, eds. San Diego: Academic Press/Elsevier, 2002. p. 867.

503. Opp, M. R., and L. A. Toth, Neural-immune interactions in the regulation of sleep. *Front Biosci*, 2003. 8: pp. 768–79.

504. Hogan, D., J. D. Morrow, E. M. Smith, and M. R. Opp, Interleukin-6 alters sleep of rats. *J Neuroimmunol*, 2003. 137(1–2): pp. 59–66.

505. Garcia-Garcia, F., H. Yoshida, and J. M. Krueger, Interleukin-8 promotes non-rapid eye movement sleep in rabbits and rats. *J Sleep Res*, 2004. 13(1): pp. 55–61.

506. Konsman, J. P., P. Parnet, and R. Dantzer, Cytokine-induced sickness behavior: mechanisms and implications. *Trends Neurosci*, 2002. 25(3): pp. 154–9.

507. Chang, F. C., and M. R. Opp, Corticotropin-releasing hormone (CRH) as a regulator of waking. *Neurosci Biobehav Rev*, 2001. 25(5): pp. 445–53.

508. Opp, M. R., and L. Imeri, Rat strains that differ in corticotropin-releasing hormone production exhibit different sleep-wake responses to interleukin 1. *Neuroendocrinology*, 2001. 73(4): pp. 272–84.

509. Imeri, L., M. Mancia, and M. R. Opp, Blockade of 5-hydroxytryptamine (serotonin)-2 receptors alters interleukin-1-induced changes in rat sleep. *Neuroscience*, 1999. 92(2): pp. 745–9.

510. Chang, F. C., and M. R. Opp, IL-1 is a mediator of increases in slow-wave sleep induced by CRH receptor blockade. *Am J Physiol Regul Integr Comp Physiol*, 2000. 279(3): pp. R793–802.

511. Krueger, J. M., and J. A. Majde, Humoral links between sleep and the immune system: research issues. *Ann N Y Acad Sci*, 2003. 992: pp. 9–20.

512. Mong, J. A., C. Krebs, and D. W. Pfaff, Perspective: micoarrays and differential display PCR-tools for studying transcript levels of genes in neuroendocrine systems. *Endocrinology*, 2002. 143(6): pp. 2002–6.

513. Krebs, C., E. Jarvis, and D. Pfaff, The 70- kDa heat shock cognate protein (Hsc73) gene is enhanced by ovarian hormones in the ventromedial hypothalamus. *Proc Natl Acad Sci U S A*, 1999. 96: pp. 1686–91.

514. Koyner, J., K. Demarest, J. McCaughran, Jr., L. Cipp, and R. Hitzemann, Identification and time dependence of quantitative trait loci for basal locomotor activity in the BXD recombinant inbred series and a B6D2 F2 intercross. *Behav Genet*, 2000. 30(3): pp. 159–70.

515. Shimomura, K., S. S. Low-Zeddies, D. P. King, T. D. Steeves, A. Whiteley, J. Kushla,

P. D. Zemenides, A. Lin, M. H. Vitaterna, G. A. Churchill, and J. S. Takahashi, Genome-wide epistatic interaction analysis reveals complex genetic determinants of circadian behavior in mice. *Genome Res*, 2001. 11(6): pp. 959–80.

516. Greenspan, R. J., E PLURIBUS UNUM, EX UNO PLURA1: quantitative and single-gene perspectives on the study of behavior. *Annu Rev Neurosci*, 2004. 27: pp. 79–105.

517. Tafti, M., B. Petit, D. Chollet, E. Neidhart, F. de Bilbao, J. Z. Kiss, P. A. Wood, and P. Franken, Deficiency in short-chain fatty acid beta-oxidation affects theta oscillations during sleep. *Nat Genet*, 2003. 34(3): pp. 320–5.

518. Pfaff, D. W., M. I. Phillips, and R. T. Rubin, *Principles of hormone/behavior relations*. San Diego: Academic Press/Elsevier, 2004.

519. Dellovade, T., J. Chan, V. B, D. Forrest, and D. Pfaff, The two thyroid hormone receptor genes have opposite effects on estrogen-stimulated sex behaviors. *Nat Neurosci*, 2000. 3: pp. 472–475.

520. Plomin, R., *Behavioral genetics*, 3rd ed. New York: W. H. Freeman, 1997.

521. Zubenko, G. S., B. Maher, H. B. Hughes, 3rd, W. N. Zubenko, J. S. Stiffler, B. B. Kaplan, and M. L. Marazita, Genome-wide linkage survey for genetic loci that influence the development of depressive disorders in families with recurrent, early-onset, major depression. *Am J Med Genet*, 2003. 123B(1): pp. 1–18.

522. Cryan, J. F., A. Markou, and I. Lucki, Assessing antidepressant activity in rodents: recent developments and future needs. *Trends Pharmacol Sci*, 2002. 23(5): pp. 238–45.

523. Fossella, J., M. I. Posner, J. Fan, J. M. Swanson, and D. W. Pfaff, Attentional phenotypes for the analysis of higher mental function. *ScientificWorldJournal*, 2002. 2: pp. 217–23.

524. Bouchard, T. J., Jr., Genetic and environmental influences on adult intelligence and special mental abilities. *Hum Biol*, 1998. 70(2): pp. 257–79.

525. Biaggioni, I., Functional anatomy of the central autonomic nervous system, in *Handbook of clinical neurology: the autonomic nervous system. Part II: Dysfunctions*, O. Appenzeller, ed. Philadelphia: Saunders, 2000.

526. Bouchard, T. J., Jr., and J. C. Loehlin, Genes, evolution, and personality. *Behav Genet*, 2001. 31(3): pp. 243–73.

527. Cowan, W. M., K. L. Kopnisky, and S. E. Hyman, The human genome project and its impact on psychiatry. *Annu Rev Neurosci*, 2002. 25: pp. 1–50.

528. Pfaff, D. W., Ogawa, S, Kia, K., Vasudevan, N., Krebs, C., Frohlich, J. and Kow, L. M., Genetic mechanisms in neural and hormonal controls over female reproductive behaviors, in *Hormones, brain and behavior*, D. W. Pfaff, A. Arnold, A. Etgen, S. Fahrbach, and R. Rubin, eds. San Diego: Academic Press/Elsevier, 2002.

529. Pfaff, D. W., Arnold, A., Etgen, A., Fahrbach, S., and R. Rubin, eds., *Hormones, brain, and behavior*, 5 vols. San Diego: Academic Press/Elsevier, 2002.

530. Wallen, K., Sex and context: hormones and primate sexual motivation. *Horm Behav*, 2001. 40(2): pp. 339–57.

531. Schmitt, A., and International Society for Human Ethology, Conference Proceedings, *New aspects of human ethology: the language of science*. New York: Plenum Press, 1997.

532. Pfaff, D., Autoradiographic localization of radioactivity in rat brain after injection of tritiated sex hormones. *Science*, 1968. 161: pp. 1355–1356.

533. Mong, J. A., and D. W. Pfaff, Hormonal symphony: steroid orchestration of gene modules for sociosexual behaviors. *Mol Psychiatry*, 2004, 9:550–6.

534. Ogawa, S., J. Taylor, D. Lubahn, K. Korach, and D. Pfaff, Reversal of sex roles in genetic female mice by disruption of estrogen receptor gene. *Neuroendocrinology*, 1996. 64: pp. 467–470.

535. Ogawa, S., V. Eng, J. Taylor, D. Lubahn, K. Korach, and D. Pfaff, Roles of estrogen re-

ceptor-alpha gene expression in reproduction-related behaviors in female mice. *Endocrinology*, 1998. 139: pp. 5070–81.

536. Ogawa, S., and D. Pfaff, Genetic contributions to the sexual differentiation of behavior, in *Sexual differentiation of the brain*, A. Matsumoto, ed. Boca Raton, Fl.: CRC Press, 2000. pp. 11–20.

537. Ogawa, S., J. Gustafsson, K. Korach, and D. Pfaff, Survival of reproduction-related behaviors in male and female estrogen receptor β deficient (βERKO) male and female mice. *Proc Natl Acad Sci U S A*, 1999. 96: pp. 12887–92.

538. Gustafsson, J. A., What pharmacologists can learn from recent advances in estrogen signalling. *Trends Pharmacol Sci*, 2003. 24(9): pp. 479–85.

539. Choleris, E., J. A. Gustafsson, K. S. Korach, L. J. Muglia, D. W. Pfaff, and S. Ogawa, An estrogen-dependent four-gene micronet regulating social recognition: a study with oxytocin and estrogen receptor-alpha and -beta knockout mice. *Proc Natl Acad Sci U S A*, 2003. 100(10): pp. 6192–7.

540. Nomura, M., Durback, L., Chan, J., Gustafsson, J-A., Smithies, O., Korach, K. S., Pfaff, D. W., and Ogawa, S., Genotype/age interactions on aggressive behavior in gonadally intact estrogen receptor β knockout (βERKO) male mice. *Horm Behav*, 2002. 41: pp. 288–96.

541. Ogawa, S., U. E. Olazabal, I. S. Parhar, and D. W. Pfaff, Effects of intrahypothalamic administration of antisense DNA for progesterone receptor mRNA on reproductive behavior and progesterone receptor immunoreactivity in female rat. *J Neurosci*, 1994. 14: pp. 1766–74.

542. Pfaff, D. W., and A. Agmo, Arousal mechanisms serving the reproductive motivation underlying reproductive behaviors, in *Steven's handbook of experimental psychology*, R. Gallistel, ed. New York: Wiley. pp. 709–36.

543. Garey, J. H., Kow, L-M, and Pfaff, D. W., Temporal and spatial quantitation of reproductive behaviors among Swiss-Webster mice in a semi-natural environment. *Horm Behav*, 2002, 42: 294–306.

544. Garey, J., Morgan, M. A., Frohlich, J., McEwen, B. S., and Pfaff, D. W., Effects of the phytoestrogen, coumestrol, on locomotor and fear-related behaviors in female mice. *Horm Behav*, 2001. 40(1): pp. 65–76.

545. Gorzalka, B. B., and I. V. Moe, Adrenal role in proceptivity and receptivity induced by two modes of estradiol treatment. *Physiol Behav*, 1994. 55(1): pp. 29–34.

546. Ogawa, S., J. Chan, J. A. Gustafsson, K. S. Korach, and D. W. Pfaff, Estrogen increases locomotor activity in mice through estrogen receptor alpha: specificity for the type of activity. *Endocrinology*, 2003. 144(1): pp. 230–9.

547. Pfaff, D. W., Neurobiological mechanisms of sexual motivation, in *The physiological mechanisms of motivation*, D. W. Pfaff, ed. New York: Springer-Verlag, 1982. pp. 287–317.

548. Pfaff, D. W., and M. Keiner, Atlas of estradiol-concentrating cells in the central nervous system of the female rat. *J Comp Neurol*, 1973. 151: pp. 121–158.

549. Fahrbach, S., J. I. Morrell, and D. W. Pfaff, Identification of medial preoptic neurons that concentrate estradiol and project to the midbrain in the rat. *J Comp Neurol*, 1986. 247: pp. 364–82.

550. Haskins, J. T., and R. L. Moss, Action of estrogen and mechanical vaginocervical stimulation on the membrane excitability of hypothalamic and midbrain neurons. *Brain Res Bull*, 1983. 10(4): pp. 489–96.

551. Sakuma, Y., Estrogen-induced changes in the neural impulse flow from the female rat preoptic region. *Horm Behav*, 1994. 28(4): pp. 438–44.

552. Takeo, T., and Y. Sakuma, Diametrically opposite effects of estrogen on the excitability

of female rat medial and lateral preoptic neurons with axons to the midbrain locomotor region. *Neurosci Res*, 1995. 22(1): pp. 73–80.

553. Hoshina, Y., T. Takeo, K. Nakano, T. Sato, and Y. Sakuma, Axon-sparing lesion of the preoptic area enhances receptivity and diminishes proceptivity among components of female rat sexual behavior. *Behav Brain Res*, 1994. 61(2): pp. 197–204.

554. Sinnamon, H. M., Glutamate and picrotoxin injections into the preoptic basal forebrain initiate locomotion in the anesthetized rat. *Brain Res*, 1987. 400(2): pp. 270–7.

555. Sinnamon, H. M., Microstimulation mapping of the basal forebrain in the anesthetized rat: the "preoptic locomotor region." *Neuroscience*, 1992. 50(1): pp. 197–207.

556. Mori, S., K. Matsuyama, J. Kohyama, Y. Kobayashi, and K. Takakusaki, Neuronal constituents of postural and locomotor control systems and their interactions in cats. *Brain Dev*, 1992. 14 Suppl: pp. S109–20.

557. Kato, A., and Y. Sakuma, Neuronal activity in female rat preoptic area associated with sexually motivated behavior. *Brain Res*, 2000. 862(1–2): pp. 90–102.

558. Wallen, K., F. Graves, and M. Wilson, Tamoxifen inhibits estrogen-induced sexual initiation in ovariectomized rhesus monkeys. Society for Neuroscience Abstracts, 2002: p. 482.13.

559. Etgen, A. M., Estrogen regulation of neurotransmitter and growth factor signaling in the brain, in *Hormones, brain and behavior*, D. W. Pfaff, A. Arnold, A.Etgen, S. Fahrbach, and R. Rubin, eds. San Diego: Academic Press/Elsevier, 2002. pp. 381–440.

560. Etgen, A. M., M. A. Ansonoff, and A. Quesada, Mechanisms of ovarian steroid regulation of norepinephrine receptor-mediated signal transduction in the hypothalamus: implications for female reproductive physiology. *Horm Behav*, 2001. 40(2): pp. 169–77.

561. Drouin, C., L. Darracq, F. Trovero, G. Blanc, J. Glowinski, S. Cotecchia, and J. P. Tassin, Alpha-1b-adrenergic receptors control locomotor and rewarding effects of psychostimulants and opiates. *J Neurosci*, 2002. 22(7): pp. 2873–84.

562. Helena, C., C. Franci, and J. Anselmo-Franci, Luteinizing hormone and luteinizing hormone-releasing hormone secretion is under locus coeruleus control in female rats. Society for Neuroscience Abstracts, 2002: p. 669.2.

563. Acosta-Martinez, M., and A. M. Etgen, Activation of mu-opioid receptors inhibits lordosis behavior in estrogen and progesterone-primed female rats. *Horm Behav*, 2002. 41(1): pp. 88–100.

564. Dohanich, G. P., D. M. McMullan, and M. M. Brazier, Cholinergic regulation of sexual behavior in female hamsters. *Physiol Behav*, 1990. 47(1): pp. 127–31.

565. Schmich, J., A. Loewke, and E. Hull, L-trans-2, 4-PDC, a glutamate reuptake inhibitor, microinjected into the MPOA, may enhance male rat sex behavior. *Horm Behav*, 2002. 41: p. 75.

566. Gulia, K. K., H. N. Mallick, and V. M. Kumar, Orexin A (hypocretin-1) application at the medial preoptic area potentiates male sexual behavior in rats. *Neuroscience*, 2003. 116(4): pp. 921–3.

567. Bancroft, J., E. Janssen, D. Strong, L. Carnes, Z. Vukadinovic, and J. S. Long, The relation between mood and sexuality in heterosexual men. *Arch Sex Behav*, 2003. 32(3): pp. 217–30.

568. Lopez, G. J., and W. C. Koller, Sexual function, dysfunction, orientation and the autonomic nervous system, in *Handbook of clinical neurology*, O. Appenzeller, ed. Philadelphia: Saunders, 2000.

569. Heritage, A., W. Stumpf, M. Sar, and L. Grant, Brainstem catecholamine neurons are target sites for sex steroid hormones. *Science*, 1980. 207: pp. 1377–9.

570. Shughrue, P., B. Komm, and I. Merchenthaler, The distribution of estrogen receptor β mRNA in the rat hypothalamus. *Steroids*, 1996. 61: pp. 678–681.

571. Shughrue, P. J., M. V. Lane, and I. Merchanthaler, Comparative distribution of estrogen receptor alpha and beta RNA in the rat central nervous system. *J Comp Neurol*, 1997. 388: pp. 507–525.

572. Haywood, S., S. Simonian, E. van der Beek, J. Bicknell, and A. Herbison, Fluctuating estrogen and progesterone receptor expression in brainstem norepinephrine neurons through the rat estrous cycle. *Endocrinology*, 1999. 140: pp. 3255–63.

573. Zhang, J. Q., W. Q. Cai, S. Zhou de, and B. Y. Su, Distribution and differences of estrogen receptor beta immunoreactivity in the brain of adult male and female rats. *Brain Res*, 2002. 935(1–2): pp. 73–80.

574. Lu, H., H. Ozawa, M. Nishi, T. Ito, and M. Kawata, Serotonergic neurones in the dorsal raphe nucleus that project into the medial preoptic area contain oestrogen receptor beta. *J Neuroendocrinol*, 2001. 13(10): pp. 839–45.

575. Creutz, L. M., and M. F. Kritzer, Estrogen receptor-beta immunoreactivity in the midbrain of adult rats: regional, subregional, and cellular localization in the A10, A9, and A8 dopamine cell groups. *J Comp Neurol*, 2002. 446(3): pp. 288–300.

576. Kelly, M. J., and O. K. Ronnekliev, Rapid membrane effects of estrogen in the central nervous system, in *Hormones, brain, and behavior*, D. W. Pfaff, A. Arnold, A.Etgen, S. Fahrbach, and R. Rubin, eds. San Diego: Academic Press/Elsevier, 2002. pp. 361–381.

577. Kelly, M. J., and E. R. Levin, Rapid actions of plasma membrane estrogen receptors. *Trends Endocrinol Metab*, 2001. 12(4): pp. 152–6.

578. Vasudevan, N. K., Kow, L. M., and Pfaff, D. W., Early membrane estrogenic effects required for full expression of slower genomic actions in a nerve cell line. *Proc Natl Acad Sci U S A*, 2001. 98: pp. 12267–71.

579. McKenna, K. E., Neural circuitry involved in sexual function. *J Spinal Cord Med*, 2001. 24(3): pp. 148–54.

580. Yanagimoto, M., K. Honda, Y. Goto, and H. Negoro, Afferents originating from the dorsal penile nerve excite oxytocin cells in the hypothalamic paraventricular nucleus of the rat. *Brain Res*, 1996. 733(2): pp. 292–6.

581. Petitti, N., G. B. Karkanias, and A. M. Etgen, Estradiol selectively regulates alpha 1B-noradrenergic receptors in the hypothalamus and preoptic area. *J Neurosci*, 1992. 12(10): pp. 3869–76.

582. Quesada, A., and A. M. Etgen, Functional interactions between estrogen and insulin-like growth factor-I in the regulation of alpha 1B-adrenoceptors and female reproductive function. *J Neurosci*, 2002. 22(6): pp. 2401–8.

583. Berridge, C. W., S. O. Isaac, and R. A. Espana, Additive wake-promoting actions of medial basal forebrain noradrenergic alpha1- and beta-receptor stimulation. *Behav Neurosci*, 2003. 117(2): pp. 350–9.

584. Osaka, T. and H. Matsumura, Noradrenergic inputs to sleep-related neurons in the preoptic area from the locus coeruleus and the ventrolateral medulla in the rat. *Neurosci Res*, 1994. 19(19): pp. 39–50.

585. Reisert, I., V. Han, E. Lieth, D. Toran-Allerand, C. Pilgrim, and J. Lauder, Sex steroids promote neurite growth in mesencephalic tyrosine hydroxylase immunoreactive neurons in vitro. *J Dev Neurosci*, 1987. 5: pp. 91–98.

586. Raab, H., C. Pilgrim, and I. Reisert, Effects of sex and estrogen on tyrosine hydroxylase mRNA in cultured embryonic rat mesencephalon. *Brain Res Mol Brain Res*, 1995. 33(1): pp. 157–64.

587. Appararundaram, S., J. Huller, S. Lakhlani, and L. Jennes, Ovariectomy-induced alterations of choline and dopamine transporter activity in the rat brain. Society for Neuroscience Abstracts, 2002: p. 368.20.

588. Auger, A. P., J. M. Meredith, G. L. Snyder, and J. D. Blaustein, Oestradiol increases

phosphorylation of a dopamine- and cyclic AMP-regulated phosphoprotein (DARPP-32) in female rat brain. *J Neuroendocrinol*, 2001. 13(9): pp. 761–8.

589. Fekete, C., P. Strutton, F. Cagampang, E. Hrabovszky, I. Kallo, P. Shughrue, E. Dobo, E. Mihaly, L. Baranyi, H. Okada, P. Panula, I. Merchenthaler, C. Coen, and Z. Liposits, Estrogen receptor immunoreactivity is present in the majority of central histaminergic neurons: evidence for a new neuroendocrine pathway associated with luteinizing hormone-releasing hormone-synthesizing neurons in rats and humans. *Endocrinology*, 1999. 140: pp. 4335–41.

590. Kow, L. M., and D. W. Pfaff, Biophysical actions of histamine on hypothalamic neurons, and their hormone sensitivity. 2005, in press.

591. Fink, G., B. E. Sumner, R. Rosie, O. Grace, and J. P. Quinn, Estrogen control of central neurotransmission: effect on mood, mental state, and memory. *Cell Mol Neurobiol*, 1996. 16(3): pp. 325–344.

592. Fink, G., B. Sumner, J. McQueen, H. Wilson, and R. Rosie, Sex steroid control of mood, mental state and memory. *Clin Exp Pharmacol Physiol*, 1998. 25: pp. 764–75.

593. Fink, G., R. Dow, J. McQueen, J. Bennie, and S. Carroll, Serotonergic 5-HT$_{2A}$ receptors important for the oestradiol-induced surge of luteinising hormone-releasing hormone in the rat. *J Neuroendocrinol*, 1999. 11: pp. 63–69.

594. Raap, D. K., L. DonCarlos, F. Garcia, N. A. Muma, W. A. Wolf, G. Battaglia, and L. D. Van de Kar, Estrogen desensitizes 5-HT(1A) receptors and reduces levels of G(z), G(i1) and G(i3) proteins in the hypothalamus. *Neuropharmacology*, 2000. 39(10): pp. 1823–32.

595. Bouali, S., A. Evrard, M. Chastanet, K. P. Lesch, M. Hamon, and J. Adrien, Sex hormone-dependent desensitization of 5-HT1A autoreceptors in knockout mice deficient in the 5-HT transporter. *Eur J Neurosci*, 2003. 18(8): pp. 2203–12.

596. Pfaff, D. W., M. McCarthy, S. Schwartz-Giblin, and L. M. Kow, Female reproductive behavior, in *The physiology of reproduction*, E. Knobil and J. Neill, eds. New York: Raven, 1994. pp. 107–220.

597. Gibbs, R. B., Effects of gonadal hormone replacement on measures of basal forebrain cholinergic function. *Neuroscience*, 2000. 101(4): pp. 931–8.

598. Gibbs, R. B., D. Wu, L. B. Hersh, and D. W. Pfaff, Effects of estrogen replacement on relative levels of choline acetyltransferase, trkA, and nerve growth factor messenger RNAs in the basal forebrain and hippocampal formation in adult rats. *Exp Neurol*, 1994. 129: pp. 70–80.

599. Saleh, T. M., and B. J. Connell, Estrogen-induced autonomic effects are mediated by NMDA and GABAA receptors in the parabrachial nucleus. *Brain Res*, 2003. 973(2): pp. 161–70.

600. Johren, O., N. Bruggemann, A. Dendorfer, and P. Dominiak, Gonadal steroids differentially regulate the messenger ribonucleic acid expression of pituitary orexin type 1 receptors and adrenal orexin type 2 receptors. *Endocrinology*, 2003. 144(4): pp. 1219–25.

601. Ceccatelli, S., L. Grandison, R. E. Scott, D. W. Pfaff, and L. M. Kow, Estradiol regulation of nitric oxide synthase mRNAs in rat hypothalamus. *Neuroendocrinology*, 1996. 64: pp. 357–63.

602. Rachman, I. M., D. W. Pfaff, and R. S. Cohen, NADPH diaphorase activity and nitric oxide synthase immunoreactivity in lordosis-relevant neurons of the ventromedial hypothalamus. *Mol Brain Res*, 1996. 740: pp. 291–306.

603. Marino, J., and J. Cudeiro, Nitric oxide-mediated cortical activation: a diffuse wake-up system. *J Neurosci*, 2003. 23(10): pp. 4299–307.

604. Corchero, J., J. Manzanares, and J. A. Fuentes, Role of gonadal steroids in the corticotropin-releasing hormone and proopiomelanocortin gene expression response

to Delta(9)-tetrahydrocannabinol in the hypothalamus of the rat. *Neuroendocrinology,* 2001. 74(3): pp. 185–92.

605. Li, H., and E. Satinoff, Body temperature and sleep in intact and ovariectomized female rats. *Am J Physiol,* 1996. 271(6 Pt 2): pp. R1753–8.

606. Colvin, G. B., D. I. Whitmoyer, R. D. Lisk, D. O. Walter, and C. H. Sawyer, Changes in sleep-wakefulness in female rats during circadian and estrous cycles. *Brain Res,* 1968. 7(2): pp. 173–81.

607. Colvin, G. B., D. I. Whitmoyer, and C. H. Sawyer, Circadian sleep-wakefulness patterns in rats after ovariectomy and treatment with estrogen. *Exp Neurol,* 1969. 25(4): pp. 616–25.

608. Manber, R., and R. Armitage, Sex, steroids, and sleep: a review. *Sleep,* 1999. 22(5): pp. 540–55.

609. Kiyatkin, E. A., and R. D. Mitchum, Jr., Fluctuations in brain temperature during sexual interaction in male rats: an approach for evaluating neural activity underlying motivated behavior. *Neuroscience,* 2003. 119(4): pp. 1169–83.

610. Mitchum, R. D., Jr., and E. A. Kiyatkin, Brain hyperthermia and temperature fluctuations during sexual interaction in female rats. *Brain Res,* 2004. 1000(1–2): pp. 110–22.

611. Rubinow, D. R., P. J. Schmidt, C. A. Roca, and R. C. Daly, Gonadal hormones and behavior in women: concentrations versus context, in *Hormones, brain, and behavior,* D. W. Pfaff, A. Arnold, A.Etgen, S. Fahrbach, and R. Rubin, eds. San Diego: Academic Press/ Elsevier, 2002. pp. 37–75.

612. Soares, J. C., and S. Gershon, Prospects for the development of new treatments with a rapid onset of action in affective disorders. *Drugs,* 1996. 52(4): pp. 477–82.

613. Marvan, M. L., L. Chavez-Chavez, and S. Santana, Clomipramine modifies fluctuations of forced swimming immobility in different phases of the rat estrous cycle. *Arch Med Res,* 1996. 27(1): pp. 83–6.

614. Marvan, M. L., S. Santana, L. Chavez Chavez, and M. Bertran, Inescapable shocks accentuate fluctuations of forced swimming immobility in different phases of the rat estrous cycle. *Arch Med Res,* 1997. 28(3): pp. 369–72.

615. Rachman, I., J. Unnerstall, D. Pfaff, and R. Cohen, Estrogen alters behavior and forebrain c-fos expression in ovariectomized rats subjected to the forced swim test. *Proc Natl Acad Sci U S A,* 1998. 95: pp. 13941–6.

616. Bernardi, M., A. V. Vergoni, M. Sandrini, S. Tagliavini, and A. Bertolini, Influence of ovariectomy, estradiol and progesterone on the behavior of mice in an experimental model of depression. *Physiol Behav,* 1989. 45(5): pp. 1067–8.

617. Blackman, C. and D. W. Pfaff. 2003, unpublished data.

618. Dietrich, T., T. Krings, J. Neulen, K. Willmes, S. Erberich, A. Thron, and W. Sturm, Effects of blood estrogen level on cortical activation patterns during cognitive activation as measured by functional MRI. *Neuroimage,* 2001. 13(3): pp. 425–32.

619. Fisher, H., A. Aron, D. Mashek, G. Strong, H. Li, and L. L. Brown, Early stage intense romantic love activates cortical—basal—ganglia reward/motivation, emotion and attention systems: an fMRI study of a dynamic network that varies with relationship length, passion intensity and gender. Society for Neuroscience Abstracts, 2003: p. 725.27.

620. Jackson, L., D. Watkins, and J. Becker, Estrogen enhances induction of behavioral sensitization to cocaine and dopamine release. Society for Neuroscience Abstracts, 2002: p. 289.2.

621. Hu, M., and J. B. Becker, Effects of sex and estrogen on behavioral sensitization to cocaine in rats. *J Neurosci,* 2003. 23(2): pp. 693–9.

622. Zhou, W., K. A. Cunningham, and M. L. Thomas, Estrogen regulation of gene expres-

sion in the brain: a possible mechanism altering the response to psychostimulants in female rats. *Brain Res Mol Brain Res*, 2002. 100(1–2): pp. 75–83.

623. Berkeley, K. J., G. E. Hoffman, A. Z. Murphy, and A. Holdcroft, Pain: sex/gender differences, in *Hormones, brain, and behavior*, D. W. Pfaff, A. Arnold, A. Etgen, S. Fahrbach, and R. Rubin, eds. San Diego: Academic Press/Elsevier, 2002. pp. 409–43.

624. Kapp, B., P. Whalen, W. Supple, and J. Pascoe, Amygdaloid contributions to conditioned arousal and sensory information processing, in *The amygdala: neurobiological aspects of emotion, memory, and mental dysfunction*, J. Aggleton, ed. 1992, New York: Whiley-Liss, 1992. pp. 51–80.

625. Davis, M., The role of the amygdala in conditioned fear, in *The amygdala: neurobiological aspects of emotion, memory, and mental dysfunction*, J. Aggleton, ed. 1992, New York: Whiley-Liss, 1992. pp. 255–306.

626. Fanselow, M., Neural organization of the defensive behavior system responsible for fear. *Psychon Rev*, 1994. 1: pp. 429–38.

627. LeDoux, J. E., *The emotional brain : the mysterious underpinnings of emotional life*. New York: Simon & Schuster, 1996.

628. Brewer, J. A., K. E. Bethin, M. L. Schaefer, L. M. Muglia, S. K. Vogt, S. C. Weninger, J. A. Majzoub, and L. J. Muglia, Dissecting adrenal and behavioral responses to stress by targeted gene inactivation in mice. *Stress*, 2003. 6(2): pp. 121–5.

629. Sternberg, E. M., *The balance within: the science connecting health and emotions*. New York: W. H. Freeman, 2000.

630. de Kloet, E. R., Hormones, brain and stress. *Endocr Regul*, 2003. 37(2): pp. 51–68.

631. de Kloet, E. R., and M. S. Oitzl, Who cares for a stressed brain? The mother, the kid or both? *Neurobiol Aging*, 2003. 24(Suppl 1): pp. S61–5; discussion S67–8.

632. Hsu, S. Y., and A. J. Hsueh, Human stresscopin and stresscopin-related peptide are selective ligands for the type 2 corticotropin-releasing hormone receptor. *Nat Med*, 2001. 7(5): pp. 605–11.

633. Smith, G. W., J. M. Aubry, F. Dellu, A. Contarino, L. M. Bilezikjian, L. H. Gold, R. Chen, Y. Marchuk, C. Hauser, C. A. Bentley, P. E. Sawchenko, G. F. Koob, W. Vale, and K. F. Lee, Corticotropin releasing factor receptor 1-deficient mice display decreased anxiety, impaired stress response, and aberrant neuroendocrine development. *Neuron*, 1998. 20(6): pp. 1093–102.

634. Bale, T. L., A. Contarino, G. W. Smith, R. Chan, L. H. Gold, P. E. Sawchenko, G. F. Koob, W. W. Vale, and K. F. Lee, Mice deficient for corticotropin-releasing hormone receptor-2 display anxiety-like behavior and are hypersensitive to stress. *Nat Genet*, 2000. 24(4): pp. 410–4.

635. Shankman, S. A., and D. N. Klein, The relation between depression and anxiety: an evaluation of the tripartite, approach-withdrawal and valence-arousal models. *Clin Psychol Rev*, 2003. 23(4): pp. 605–37.

636. Morgan, M. A., J. Schulkin, and D. W. Pfaff, Estrogens and non-reproductive behaviors related to activity and fear. *Neurosci Biobehav Rev*, 2004. 28(1): pp. 55–63.

637. Westberg, L., J. Melke, M. Landen, S. Nilsson, F. Baghaei, R. Rosmond, M. Jansson, G. Holm, P. Bjorntorp, and E. Eriksson, Association between a dinucleotide repeat polymorphism of the estrogen receptor alpha gene and personality traits in women. *Mol Psychiatry*, 2003. 8(1): pp. 118–22.

638. Maier, S. F., R. C. Drugan, and J. W. Grau, Controllability, coping behavior and stress-induced analgesia in the rat. *Pain*, 1982. 12: pp. 47–56.

639. Telner, J. I., Z. Merali, and R. L. Singhal, Stress controllability and plasma prolactin levels in the rat. *Psychoneuroendocrinology*, 1982. 7(4): pp. 361–4.

640. Haracz, J. L., T. R. Minor, J. N. Wilkins, and E. G. Zimmermann, Learned helplessness: an experimental model of the DST in rats. *Biol Psychiatry*, 1988. 23(4): pp. 388–96.

641. Weiss, J. M., Behavioral depression produced by an uncontrollable stressor: relationship to norepinephrine, dopamine, and serotonin levels in various regions of rat brain. *Brain Res Revs*, 1981. 3: pp. 167–205.

642. Bland, S. T., C. Twining, L. R. Watkins, and S. F. Maier, Stressor controllability modulates stress-induced serotonin but not dopamine efflux in the nucleus accumbens shell. *Synapse*, 2003. 49(3): pp. 206–8.

643. Shors, T. J., T. B. Seib, S. Levine, and R. F. Thompson, Inescapable versus escapable shock modulates long-term potentiation in the rat hippocampus. *Science*, 1989. 244(4901): pp. 224–6.

644. Seligman, M. E., J. Weiss, M. Weinraub, and A. Schulman, Coping behavior: learned helplessness, physiological change and learned inactivity. *Behav Res Ther*, 1980. 18(5): pp. 459–512.

645. Helmreich, D. L., L. R. Watkins, T. Deak, S. F. Maier, H. Akil, and S. J. Watson, The effect of stressor controllability on stress-induced neuropeptide mRNA expression within the paraventricular nucleus of the hypothalamus. *J Neuroendocrinol*, 1999. 11(2): pp. 121–8.

646. Qin, Y., H. Karst, and M. Joels, Chronic unpredictable stress alters gene expression in rat single dentate granule cells. *J Neurochem*, 2004. 89(2): pp. 364–74.

647. Jenkins, W. J., and J. B. Becker, Dynamic increases in dopamine during paced copulation in the female rat. *Eur J Neurosci*, 2003. 18(7): pp. 1997–2001.

648. Plutchik, R., *The psychology and biology of emotion*, 1st ed. New York: HarperCollins College Publishers, 1994.

649. Moss, R., and S. McCann, Induction of mating behavior in rats by luteinizing hormone-releasing factor. *Science*, 1973. 181: pp. 177–9.

650. Berridge, K. C., Comparing the emotional brains of humans and other animals, in *Handbook of affective sciences*, R. J. Davidson, K. R. Scherer, and H. H. Goldsmith, eds. Oxford: Oxford University Press, 2003. pp. 25–51.

651. Schulkin, J., *Rethinking homeostasis: allostatic regulation in physiology and pathophysiology*. Cambridge, Mass.: The MIT Press, 2003.

652. Cacioppo, J. T., M. H. Burleson, K. M. Poehlmann, W. B. Malarkey, J. K. Kiecolt-Glaser, G. G. Berntson, B. N. Uchino, and R. Glaser, Autonomic and neuroendocrine responses to mild psychological stressors: effects of chronic stress on older women. *Ann Behav Med*, 2000. 22(2): pp. 140–8.

653. Mayer, E. A., The neurobiology of stress and gastrointestinal disease. *Gut*, 2000. 47(6): pp. 861–9.

654. Lee, O. Y., E. A. Mayer, M. Schmulson, L. Chang, and B. Naliboff, Gender-related differences in IBS symptoms. *Am J Gastroenterol*, 2001. 96(7): pp. 2184–93.

655. McEwen, B. S., and E. N. Lasley, The end of stress as we know it. Washington, D.C.: Joseph Henry Press, 2002.

656. Fisher, H. E., *Why we love: the nature and chemistry of romantic love*, 1st ed. New York: Henry Holt and Company: 2004.

657. Arnow, B. A., J. E. Desmond, L. L. Banner, G. H. Glover, A. Solomon, M. L. Polan, T. F. Lue, and S. W. Atlas, Brain activation and sexual arousal in healthy, heterosexual males. *Brain*, 2002. 125(Pt 5): pp. 1014–23.

658. Holstege, G., A. Reinders, A. Paans, L. Meiners, J. Pruims, and J. Georgiadis, Brain activation during female sexual orgasm. Society for Neuroscience Abstracts, 2003: p. 727.7.

659. Jankowska, E., I. Hammar, U. Slawinska, K. Maleszak, and S. A. Edgley, Neuronal basis of crossed actions from the reticular formation on feline hindlimb motoneurons. *J Neurosci*, 2003. 23(5): pp. 1867–78.

660. Kushida, C. A., D. B. Rye, D. Nummy, J. G. Milton, J. P. Spire, and A. Rechtschaffen, Cortical asymmetry of REM sleep EEG following unilateral pontine hemorrhage. *Neurology*, 1991. 41(4): pp. 598–601.

661. Viau, V. and P. E. Sawchenko, Hypophysiotropic neurons of the paraventricular nucleus respond in spatially, temporally, and phenotypically differentiated manners to acute vs. repeated restraint stress: rapid publication. *J Comp Neurol*, 2002. 445(4): pp. 293–307.

662. Sawchenko, P. E., E. R. Brown, R. K. Chan, A. Ericsson, H. Y. Li, B. L. Roland, and K. J. Kovacs, The paraventricular nucleus of the hypothalamus and the functional neuroanatomy of visceromotor responses to stress. *Prog Brain Res*, 1996. 107: pp. 201–22.

663. Kc, P., M. A. Haxhiu, F. P. Tolentino-Silva, M. Wu, C. O. Trouth, and S. O. Mack, Paraventricular vasopressin-containing neurons project to brain stem and spinal cord respiratory-related sites. *Respir Physiol Neurobiol*, 2002. 133(1–2): pp. 75–88.

664. Gaus, S. E., R. E. Strecker, B. A. Tate, R. A. Parker, and C. B. Saper, Ventrolateral preoptic nucleus contains sleep-active, galaninergic neurons in multiple mammalian species. *Neuroscience*, 2002. 115(1): pp. 285–94.

665. Chou, T. C., A. A. Bjorkum, S. E. Gaus, J. Lu, T. E. Scammell, and C. B. Saper, Afferents to the ventrolateral preoptic nucleus. *J Neurosci*, 2002. 22(3): pp. 977–90.

666. Alam, M. N., R. Szymusiak, H. Gong, J. King, and D. McGinty, Adenosinergic modulation of rat basal forebrain neurons during sleep and waking: neuronal recording with microdialysis. *J Physiol*, 1999. 521(Pt 3): pp. 679–90.

667. Methippara, M. M., M. N. Alam, R. Szymusiak, and D. McGinty, Effects of lateral preoptic area application of orexin-A on sleep-wakefulness. *Neuroreport*, 2000. 11(16): pp. 3423–6.

668. McGinty, D., and R. Szymusiak, The sleep-wake switch: a neuronal alarm clock. *Nat Med*, 2000. 6(5): pp. 510–1.

669. Steininger, T. L., H. Gong, D. McGinty, and R. Szymusiak, Subregional organization of preoptic area/anterior hypothalamic projections to arousal-related monoaminergic cell groups. *J Comp Neurol*, 2001. 429(4): pp. 638–53.

670. Westerhaus, M. J., and A. D. Loewy, Sympathetic-related neurons in the preoptic region of the rat identified by viral transneuronal labeling. *J Comp Neurol*, 1999. 414(3): pp. 361–78.

671. Loewy, A. D., Forebrain nuclei involved in autonomic control. *Prog Brain Res*, 1991. 87: pp. 253–68.

672. Condes-Lara, M., N. M. Gonzalez, G. Martinez-Lorenzana, O. L. Delgado, and M. J. Freund-Mercier, Actions of oxytocin and interactions with glutamate on spontaneous and evoked dorsal spinal cord neuronal activities. *Brain Res*, 2003. 976(1): pp. 75–81.

673. Schulkin, J., B. L. Thompson, and J. B. Rosen, Demythologizing the emotions: adaptation, cognition, and visceral representations of emotion in the nervous system. *Brain Cogn*, 2003. 52(1): pp. 15–23.

674. Getting, P. A., Emerging principles governing the operation of neural networks. *Annu Rev Neurosci*, 1989. 12: pp. 185–204.

675. Li, Y., X. B. Gao, T. Sakurai, and A. N. van den Pol, Hypocretin/orexin excites hypocretin neurons via a local glutamate neuron-A potential mechanism for orchestrating the hypothalamic arousal system. *Neuron*, 2002. 36(6): pp. 1169–81.

676. Wu, M., L. Zaborszky, T. Hajszan, A. N. van den Pol, and M. Alreja, Hypocretin/

orexin innervation and excitation of identified septohippocampal cholinergic neurons. *J Neurosci*, 2004. 24(14): pp. 3527–36.

677. Valentino, R., M. Page, and A. Curtis, Activation of noradrenergic locus coeruleus neurons by hemodynamic stress is due to local release of corticotropin-releasing factor. *Brain Res*, 1991. 555: pp. 25–34.

678. Valentino, R. J., Corticotropin-releasing factor: putative neurotransmitter in the noradrenergic nucleus locus coeruleus. *Psychopharmacol Bull*, 1989. 25(3): pp. 306–11.

679. Conti, L. H., and S. L. Foote, Effects of pretreatment with corticotropin-releasing factor on the electrophysiological responsivity of the locus coeruleus to subsequent corticotropin-releasing factor challenge. *Neuroscience*, 1995. 69(1): pp. 209–19.

680. Vacher, C. M., P. Fretier, C. Creminon, A. Calas, and H. Hardin-Pouzet, Activation by serotonin and noradrenaline of vasopressin and oxytocin expression in the mouse paraventricular and supraoptic nuclei. *J Neurosci*, 2002. 22(5): pp. 1513–22.

681. Milton, J. G., A. Longtin, A. Beuter, M. C. Mackey, and L. Glass, Complex dynamics and bifurcations in neurology. *J Theor Biol*, 1989. 138(2): pp. 129–47.

682. Lewis, F. L., *Applied optimal control and estimation: digital design and implementation.* Prentice Hall and Texas Instruments Digital Signal Processing Series. Englewood Cliffs, N.J.: Prentice Hall, 1992.

683. Milsum, J. H., *Biological control systems analysis.* McGraw-Hill Electronic Science Series. New York: McGraw-Hill, 1966.

684. Germain, R. N., The art of the probable: system control in the adaptive immune system. *Science*, 2001. 293(5528): pp. 240–5.

685. Stewart, I., Networking opportunity. *Nature*, 2004. 427(6975): pp. 601–4.

686. Ogawa, S., E. Choleris, and D. W. Pfaff, Genes affecting aggressive behaviors in animals, in *Molecular mechanisms underlying aggression*, G. Bock, ed. of Pfaff, Nelson & Keverne, Novartis Symposium. New York: Wiley, in press.

687. Aleksander, I., and F. K. Hanna, *Automata theory: an engineering approach.* New York: Crane Russak, 1976.

688. Khoussainov, B., and A. Nerode, *Automata theory and its applications: progress in computer science and applied logic*, vol. 21. Boston: Birkhauser, 2001.

689. Hopcroft, J. E., and J. D. Ullman, *Introduction to automata theory, languages, and computation.* Reading, Mass.: Addison-Wesley, 1979.

690. Gill, A., *Introduction to the theory of finite-state machines.* McGraw-Hill Electronic Sciences Series. New York: McGraw-Hill, 1962.

691. Mealy, G. H., A method for synthesizing sequential circuits. *Bell Sys Tech J*, 1955. 34: pp. 1045–1079.

692. Bavel, Z., *Introduction to the theory of automata.* Reston, Va.: Reston Publishing, 1983.

693. Jansson, C., Rigorous lower and upper bounds in linear programming. *SIAM Journal on Optimization*, 2004. 14(13): pp. 914–35.

694. Kolda, T. G., and V. J. Torczon, On the convergence of asynchronous parallel pattern search. *SIAM Journal on Optimization*, 2004. 14(4): pp. 939–64.

695. Pfaff, D. W., and E. Gregory, Correlation between preoptic area unit activity and the cortical EEG: difference between normal and castrated male rats. *Electroenceph Clin Neurophysiol*, 1971. 31: pp. 223–230.

696. McCulloch, W. S., and W. Pitts, A logical calculus of the ideas immanent in nervous activity. Mathematical Biophysics, 1943. 5(115–133): pp. 99–115; discussion 73–97.

697. McCulloch, W. S., Introduction to the problem of the reticular formation, in *Automata theory*, E. R. Caianiello, ed. New York: Academic Press, 1966.

698. Merriam, C. W., *Optimization theory and the design of feedback control systems.* McGraw-Hill Electronic Sciences Series. New York: McGraw-Hill, 1964.

699. Sontag, E. D., *Mathematical control theory: deterministic finite dimensional systems*, 2nd ed. New York: Springer, 1998.

700. Cowan, J. D., Synthesis of reliable automata from unreliable components, in *Automata theory*, E. R. Caianiello, ed. New York: Academic Press, 1966.

701. Shannon, C. E., J. McCarthy, and W. R. Ashby, *Automata studies*. Annals of Mathematics Studies, no. 34. Princeton: Princeton University Press, 1956.

702. Jeong, H., S. P. Mason, A. L. Barabasi, and Z. N. Oltvai, Lethality and centrality in protein networks. *Nature*, 2001. 411(6833): pp. 41–2.

703. von Baeyer, H. C., *Information: the new language of science*. Cambridge, Mass.: Harvard University Press, 2004.

704. Sato, K., Y. Ito, T. Yomo, and K. Kaneko, On the relation between fluctuation and response in biological systems. *Proc Natl Acad Sci U S A*, 2003. 100(24): pp. 14086–90.

705. Reeke, G. N., and A. D. Coop, Estimating the temporal interval entropy of neuronal discharge. *Neural Comput*, 2004. 16(5): pp. 941–70.

706. Ueyama, T., K. E. Krout, X. V. Nguyen, V. Karpitskiy, A. Kollert, T. C. Mettenleiter, and A. D. Loewy, Suprachiasmatic nucleus: a central autonomic clock. *Nat Neurosci*, 1999. 2(12): pp. 1051–3.

707. Reick, M., J. A. Garcia, C. Dudley, and S. L. McKnight, NPAS2: an analog of clock operative in the mammalian forebrain. *Science*, 2001. 293(5529): pp. 506–9.

708. Terazono, H., T. Mutoh, S. Yamaguchi, M. Kobayashi, M. Akiyama, R. Udo, S. Ohdo, H. Okamura, and S. Shibata, Adrenergic regulation of clock gene expression in mouse liver. *Proc Natl Acad Sci U S A*, 2003. 100(11): pp. 6795–800.

708a. Dijk, D. J., and C. A. Czeisler, Contribution of the circadian pacemaker and sleep homeostat to sleep propensity. *J Neurosci*, 1995. 15(5): pp. 3526–3538.

708b. Marchant, E. G., and R. E. Mistlberger, Adaptation to feeding time in intact and SCN-ablated mice. *Brain Research*. 1997. 765: pp. 273–282.

709. Buijs, R. M., and A. Kalsbeek, Hypothalamic integration of central and peripheral clocks. *Nat Rev Neurosci*, 2001. 2(7): pp. 521–6.

710. Lu, J., Y. H. Zhang, T. C. Chou, S. E. Gaus, J. K. Elmquist, P. Shiromani, and C. B. Saper, Contrasting effects of ibotenate lesions of the paraventricular nucleus and sub-paraventricular zone on sleep-wake cycle and temperature regulation. *J Neurosci*, 2001. 21(13): pp. 4864–74.

711. Usher, M., J. D. Cohen, D. Servan-Schreiber, J. Rajkowski, and G. Aston-Jones, The role of locus coeruleus in the regulation of cognitive performance. *Science*, 1999. 283(5401): pp. 549–54.

712. Kensinger, E. A., and S. Corkin, Two routes to emotional memory: distinct neural processes for valence and arousal. *Proc Natl Acad Sci U S A*, 2004. 101(9): pp. 3310–5.

713. Davidson, R. J., Cognitive neuroscience needs affective neuroscience (and vice versa). *Brain Cogn*, 2000. 42(1): pp. 89–92.

714. Gray, J. R., T. S. Braver, and M. E. Raichle, Integration of emotion and cognition in the lateral prefrontal cortex. *Proc Natl Acad Sci U S A*, 2002. 99(6): pp. 4115–20.

715. Sanfey, A. G., J. K. Rilling, J. A. Aronson, L. E. Nystrom, and J. D. Cohen, The neural basis of economic decision-making in the Ultimatum Game. *Science*, 2003. 300(5626): pp. 1755–8.

716. Teves, D., T. O. Videen, P. E. Cryer, and W. J. Powers, Activation of human medial prefrontal cortex during autonomic responses to hypoglycemia. *Proc Natl Acad Sci U S A*, 2004. 101(16): pp. 6217–21.

717. Davidson, R. J., The neuroscience of affective style, in *The new cognitive neurosciences*, M. S. Gazzaniga, ed. Cambridge, Mass.: The MIT Press, 2000.

718. Plomin, R., J. Defries, G. McClearn, and P. McGuffin, *Behavioral genetics*, 4th ed. New York: Worth Pubishers, 2001.

719. Allport, G. W., Traits revisited. *Am Psychol*, 1966. 21: pp. 1–10.

720. Posner, M. I. and M. K. Rothbart, Attention, self-regulation and consciousness. *Philos Trans R Soc Lond B Biol Sci*, 1998. 353(1377): pp. 1915–27.

721. Burleson, M. H., K. M. Poehlmann, L. C. Hawkley, J. M. Ernst, G. G. Berntson, W. B. Malarkey, J. K. Kiecolt-Glaser, R. Glaser, and J. T. Cacioppo, Neuroendocrine and cardiovascular reactivity to stress in mid-aged and older women: long-term temporal consistency of individual differences. *Psychophysiology*, 2003. 40(3): pp. 358–69.

722. Schneidman, E., S. Still, M. J. Berry, II, and W. Bialek, Network information and connected correlations. *Phys Rev Lett*, 2003. 91(23): pp. 87–101.

723. Giacino, J. T., and J. Whyte, Amantadine to improve neurorecovery in traumatic brain injury-associated diffuse axonal injury: a pilot double-blind randomized trial. *J Head Trauma Rehabil*, 2003. 18(1): pp. 4–5; author reply 5–6.

724. Schiff, N. D., U. Ribary, D. R. Moreno, B. Beattie, E. Kronberg, R. Blasberg, J. Giacino, C. McCagg, J. J. Fins, R. Llinas, and F. Plum, Residual cerebral activity and behavioral fragments can remain in the persistently vegetative brain. *Brain*, 2002. 125(Pt 6): pp. 1210–34.

725. Giacino, J. T., S. Ashwal, N. Childs, R. Cranford, B. Jennett, D. I. Katz, J. P. Kelly, J. H. Rosenberg, J. Whyte, R. D. Zafonte, and N. D. Zasler, The minimally conscious state: definition and diagnostic criteria. *Neurology*, 2002. 58(3): pp. 349–53.

726. Hobson, J. A., The dream drugstore: chemically altered states of consciousness. Cambridge, Mass.: The MIT Press, 2001.

727. Gan, W. B., E. Kwon, G. Feng, J. R. Sanes, and J. W. Lichtman, Synaptic dynamism measured over minutes to months: age-dependent decline in an autonomic ganglion. *Nat Neurosci*, 2003. 6(9): pp. 956–60.

728. Arseven, A., J. M. Guralnik, E. O'Brien, K. Liu, and M. M. McDermott, Peripheral arterial disease and depressed mood in older men and women. *Vasc Med*, 2001. 6(4): pp. 229–34.

729. Cotman, C. W., and N. C. Berchtold, Exercise: a behavioral intervention to enhance brain health and plasticity. *Trends Neurosci*, 2002. 25(6): pp. 295–301.

730. Jennum, P., A. M. Drewes, A. Andreasen, and K. D. Nielsen, Sleep and other symptoms in primary fibromyalgia and in healthy controls. *J Rheumatol*, 1993. 20(10): pp. 1756–9.

731. Demitrack, M. A., J. K. Dale, S. E. Straus, L. Laue, S. J. Listwak, M. J. Kruesi, G. P. Chrousos, and P. W. Gold, Evidence for impaired activation of the hypothalamic-pituitary-adrenal axis in patients with chronic fatigue syndrome. *J Clin Endocrinol Metab*, 1991. 73(6): pp. 1224–34.

732. Scott, L. V., S. Medbak, and T. G. Dinan, Blunted adrenocorticotropin and cortisol responses to corticotropin-releasing hormone stimulation in chronic fatigue syndrome. *Acta Psychiatr Scand*, 1998. 97(6): pp. 450–7.

733. Scott, L. V., S. Medbak, and T. G. Dinan, The low dose ACTH test in chronic fatigue syndrome and in health. *Clin Endocrinol (Oxf)*, 1998. 48(6): pp. 733–7.

734. Scott, L. V., and T. G. Dinan, The neuroendocrinology of chronic fatigue syndrome: focus on the hypothalamic-pituitary-adrenal axis. *Funct Neurol*, 1999. 14(1): pp. 3–11.

735. Cleare, A. J., J. Bearn, T. Allain, A. McGregor, S. Wessely, R. M. Murray, and V. O'Keane, Contrasting neuroendocrine responses in depression and chronic fatigue syndrome. *J Affect Disord*, 1995. 34(4): pp. 283–9.

736. Swaab, D. F., E. J. Dubelaar, M. A. Hofman, E. J. Scherder, E. J. van Someren, and

R. W. Verwer, Brain aging and Alzheimer's disease; use it or lose it. *Prog Brain Res*, 2002. 138: pp. 343–73.

737. Hoogendijk, W. J., C. W. Pool, D. Troost, E. van Zwieten, and D. F. Swaab, Image analyser-assisted morphometry of the locus coeruleus in Alzheimer's disease, Parkinson's disease and amyotrophic lateral sclerosis. *Brain*, 1995. 118(Pt 1): pp. 131–43.

738. Iversen, L. L., M. N. Rossor, G. P. Reynolds, R. Hills, M. Roth, C. Q. Mountjoy, S. L. Foote, J. H. Morrison, and F. E. Bloom, Loss of pigmented dopamine-beta-hydroxylase positive cells from locus coeruleus in senile dementia of Alzheimer's type. *Neurosci Lett*, 1983. 39(1): pp. 95–100.

739. Raadsheer, F. C., J. J. van Heerikhuize, P. J. Lucassen, W. J. Hoogendijk, F. J. Tilders, and D. F. Swaab, Corticotropin-releasing hormone mRNA levels in the paraventricular nucleus of patients with Alzheimer's disease and depression. *Am J Psychiatry*, 1995. 152(9): pp. 1372–6.

740. Salehi, A., S. Heyn, N. K. Gonatas, and D. F. Swaab, Decreased protein synthetic activity of the hypothalamic tuberomamillary nucleus in Alzheimer's disease as suggested by smaller Golgi apparatus. *Neurosci Lett*, 1995. 193(1): pp. 29–32.

741. Quigley, K. S., L. F. Barrett, and S. Weinstein, Cardiovascular patterns associated with threat and challenge appraisals: a within-subjects analysis. *Psychophysiology*, 2002. 39(3): pp. 292–302.

742. Lin, J. S., Y. Hou, and M. Jouvet, Potential brain neuronal targets for amphetamine-, methylphenidate-, and modafinil-induced wakefulness, evidenced by c-fos immunocytochemistry in the cat. *Proc Natl Acad Sci U S A*, 1996. 93(24): pp. 14128–33.

743. Kryger, M. H., T. Roth, and W. C. Dement, *Principles and practice of sleep medicine*, 3rd ed. Philadelphia: Saunders, 2000.

744. Daan, S., D. G. Beersma, and A. A. Borbely, Timing of human sleep: recovery process gated by a circadian pacemaker. *Am J Physiol*, 1984. 246(2 Pt 2): pp. R161–83.

745. Caspi, A., J. McClay, T. E. Moffitt, J. Mill, J. Martin, I. W. Craig, A. Taylor, and R. Poulton, Role of genotype in the cycle of violence in maltreated children. *Science*, 2002. 297(5582): pp. 851–4.

746. Bock, G., ed. of D. W. Pfaff, R. Nelson, and B. Keverne, *Molecular mechanisms underlying aggression*. New York: Wiley, 2005, in press.

747. Devine, J., J. Gilligan, K. Mizcek, and D. W. Pfaff, eds. *Scientific approaches to the prevention of youth violence*. Annals, New York Academy of *Sciences*, vol. 1036, 2004.

748. Guyton, A. C., and J. E. Hall, *Textbook of medical physiology*, 9th ed. Philadelphia: Saunders, 1996.

749. Bjorntorp, P., Hypothalamic origin of prevalent human disease, in *Hormones, brain, and behavior*, D. W. Pfaff, A. Arnold, A.Etgen, S. Fahrbach, and R. Rubin, eds. San Diego: Academic Press/Elsevier, 2002. pp. 607–36.

750. Cryan, J. F., O. F. O'Leary, S. H. Jin, J. C. Friedland, M. Ouyang, B. R. Hirsch, M. E. Page, A. Dalvi, S. A. Thomas, and I. Lucki, Norepinephrine-deficient mice lack responses to antidepressant drugs, including selective serotonin reuptake inhibitors. *Proc Natl Acad Sci U S A*, 2004. 101(21): pp. 8186–91.

751. Rubin, R. T., S. M. O'Toole, M. E. Rhodes, L. K. Sekula, and R. K. Czambel, Hypothalamo-pituitary-adrenal cortical responses to low-dose physostigmine and arginine vasopressin administration: sex differences between major depressives and matched control subjects. *Psychiatry Res*, 1999. 89(1): pp. 1–20.

752. Rhodes, M. E., S. M. O'Toole, S. L. Wright, R. K. Czambel, and R. T. Rubin, Sexual diergism in rat hypothalamic-pituitary-adrenal axis responses to cholinergic stimulation and antagonism. *Brain Res Bull*, 2001. 54(1): pp. 101–13.

753. Rhodes, M. E., S. M. O'Toole, R. K. Czambel, and R. T. Rubin, Male-female dif-

ferences in rat hypothalamic-pituitary-adrenal axis responses to nicotine stimulation. *Brain Res Bull*, 2001. 54(6): pp. 681–8.

754. Stanski, D. R., Monitoring depth of anesthesia, in *Anesthesia*, R. D. Miller, ed. New York: Churchill Livingstone, 1994. pp. 1127–59.

755. Moss, J., and P. A. Craigo, The autonomic nervous system, in *Anesthesia*, R. D. Miller, ed. New York: Churchill Livingstone, 1994. pp. 523–69.

756. Reynolds, D. S., T. W. Rosahl, J. Cirone, G. F. O'Meara, A. Haythornthwaite, R. J. Newman, J. Myers, C. Sur, O. Howell, A. R. Rutter, J. Atack, A. J. Macaulay, K. L. Hadingham, P. H. Hutson, D. Belelli, J. J. Lambert, G. R. Dawson, R. McKernan, P. J. Whiting, and K. A. Wafford, Sedation and anesthesia mediated by distinct GABA(A) receptor isoforms. *J Neurosci*, 2003. 23(24): pp. 8608–17.

757. Gan, T. J., P. S. Glass, J. Sigl, P. Sebel, F. Payne, C. Rosow, and P. Embree, Women emerge from general anesthesia with propofol/alfentanil/nitrous oxide faster than men. *Anesthesiology*, 1999. 90(5): pp. 1283–7.

Acknowledgments

What a pleasure it is to acknowledge several mentors and colleagues whose teaching, advice and criticisms have helped lead to and then improve this book.

At M.I.T., Professor Hans-Lukas Teuber introduced us to ethology, the science of animal behavior in natural settings. This book can be seen as moving beyond ethology, helping to turn it into a quantitative science and explaining mechanisms of the most fundamental ethological concept. During those same graduate student years, Professor Walle J. H. Nauta elucidated the neuro-anatomic bases then understood for animal and human behaviors, inspiring an entire generation in the Department of Brain and Cognitive Sciences.

I am grateful to several colleagues and friends for reading a draft of this book and generously offering their corrections. These include Steve Henricksen (Scripps Institute), Gary Aston-Jones (University of Pennsylvania), Helen Fisher (Rutgers), Carol Lamberg, and Kevin McKenna (Northwestern University).

For Chapter 1, Kate Finley, during her summer reading, uncovered some of the examples of interactions among arousal (drive) states. At Rockefeller, Professor George Reeke and Postdoctoral Associate Alan Coop are providing both the background and the collaboration needed to apply information theoretic concepts to neuronal electrical activity and behavior. Professor Joel Cohen was very helpful regarding the presentation of an equation that explains arousal of brain and behavior as a compound increasing function of generalized CNS arousal and many specific CNS states.

For Chapter 2, Professor Bruce Kapp (University of Vermont) supplied preprints of useful material. For Chapter 3, I am delighted to thank Professor Peggy Mason (University of Chicago) and Professor Jon Horvitz (Boston College) both for illustrations of their results and many enlightening discussions over the years.

For Chapter 4, Dr. Maria Morgan's experimental work while she was at Rockefeller and her literature reviews showed how hormone-heightened arousal states could play into sex behavior under safe conditions, but into fear responses under scary conditions. Her new advisor Jay Schulkin has discussed the fundamental contrast of approach versus avoidance responses, for example, sex versus fear. For Chapter 8, my understanding of comatose, vegetative, and minimally conscious states has depended on the writing and conversation of Professor Nicholas Schiff (Cornell University School of Medicine).

At Harvard University Press, my editor Michael Fisher has been a continuing source of good advice. The warnings about the difficulties of this book's attempt, from Steve Hyman (now Provost at Harvard), were well taken. I realize that, even as this book reviews well-established bodies of data, the joining of fields theoretically proposed here—from the mathematics of information theory, to neuroanatomy and neurophysiology, to genetics, to animal and human behavior—is not without its hazards.

Index

nonspecific, 30
noradrenergic projections to, 32
visual signals to, 29, 65–66
Thermodynamics, 138–139
Thyroid hormone receptors, 93
Timing, 136–137
Tinbergen, N., 3, 12
Toxic states, 150
Trap states, 134
Tryptophan hydroxylase, 85
Tuberomammillary nucleus, 37, 41
Tyrosine hydroxylase, 84

Uncertainty
human behavior and, 139
information content and, 13–14, 15,
19–21
neural response and, 66–68
sexual arousal and, 120
Unpredictability, 14, 41–42
in ascending systems, 41–42
fear and, 119
information content and, 14, 15, 19–21
neural response and, 66–68

Utility, expected, 66

Vale, W., 116
Valentino, R. J., 130
Vegetative state, 37, 53, 147–148
Ventral tegmental area, 34, 50, 90, 113
Ventrolateral preoptic area, 40
Ventromedial nucleus, 69
Vestibular stimuli, 56
Vigilance, 150
Violence, 151
Vision, 29–30, 65–66
Von Baeyer, H. C., 138, 139
Von Holst, E., 2

Wall, P., 42
Wallen, K., 106
Weil, A., 93
Willard, J., 138
Wind up phenomena, 130

Yanagisawa, M., 90

Zaborsky, L., 30, 86, 90